Send The Stumps Flying

The Science of Fast Bowling

Send The Stumps Flying

The Science of Fast Bowling

Edited by
BRUCE ELLIOTT
DARYL FOSTER
BRIAN BLANKSBY

Foreword by
D. K. LILLEE

UNIVERSITY OF WESTERN AUSTRALIA PRESS

First published in 1989
by the University of Western Australia Press
Nedlands, W.A. 6009

Agents: Eastern States of Australia, New Zealand and Papua New Guinea: Melbourne University Press, Carlton South, Vic. 3053; U.K., Europe, Africa and Middle East: Peter Moore, P.O. Box 66, 200a Perne Road, Cambridge CB1 3PD, England; U.S.A., Canada and the Caribbean: International Specialized Book Services Inc., 5602 N.E. Hassalo Street, Portland, Oregon 97213, U.S.A.; Singapore and Malaysia: National University of Singapore Multi-Purpose Co-operative Society Ltd, Ground Floor, Central Library, Kent Ridge, Singapore 0511.

© Department of Human Movement and Recreation Studies
The University of Western Australia 1989

National Library of Australia
Cataloguing-in-Publication data

Send the stumps flying.

Bibliography.
Includes index.
ISBN 0 85564 305 6.

1. Cricket—Bowling. I. Elliott, Bruce (Bruce C.).
II. Foster, Daryl, 1939- . III. Blanksby, B. A. (Brian A.), 1941- .

796.35'822

Designed and phototypeset by the University of Western Australia Press, printed and bound by Silex Enterprise & Printing Co., Hong Kong

Contents

Foreword: D. K. Lillee   vii

The Authors   ix

CHAPTER 1: **The Fast Bowler**   *Bruce Elliott*   1

CHAPTER 2: **Fitness for Fast Bowling**   *Daryl Foster and David John*   3

Physiological Requirements of Fast Bowling
Physical Requirements of Fast Bowling
Physical and Physiological Assessment
Principles for Developing Fitness
Training Programmes for Fast Bowlers
Pre-game and Pre-training Considerations
Game Workload Requirements

CHAPTER 3: **Fast Bowling Technique**   *Bruce Elliott and Daryl Foster*   26

The Grip
The Run-up to Back-foot Landing
The Delivery Stride
Captaining the Fast Bowler

CHAPTER 4: **The Art of Swing Bowling**   *Daryl Foster and Bruce Elliott*   37

Out-swing Bowling
In-swing Bowling
Cutting and Seaming the Ball
Variation is the Key to Success

CHAPTER 5: **Psychology of Fast Bowling**   *Sandy Gordon*   43

Demands of Fast Bowlers
Mental Skills for Fast Bowlers
Communication with the Fast Bowler
Conclusion

CHAPTER 6:   **Factors that may Predispose a Fast Bowler to Injury**   54
                    Bruce Elliott, David John and Daryl Foster

             Physical Attributes
             Technique
             Physical Demands

CHAPTER 7:   **Common Injuries to the Fast Bowler**   *Ken Fitch*   60

             Injuries to the Back
             Lower-limb Injuries
             Trunk Injuries
             Upper-limb Injuries

CHAPTER 8:   **Rehabilitation of the Injured Bowler**   68
                    David John, Brian Blanksby and Sandy Gordon

             Physical Rehabilitation
             Psychological Aspects of Response to and
               Rehabilitation from Injury

References   94
Index   95

Foreword

This book covers all aspects of fast bowling, including bowling techniques and the physical and psychological preparation for play.

The authors' coaching and research experience on how to prevent injury and the correct rehabilitation procedures to follow if an injury does occur is also included. When a serious back injury threatened my cricket career I worked closely with the highly respected sport scientists in the Department of Human Movement Studies at The University of Western Australia, to rehabilitate myself and resume my fast bowling.

I am pleased to see the research findings, the sport science knowledge and the practical experience of the Western Australian Sheffield Shield Coach, Daryl Foster, made available to the general cricketing public.

I strongly recommend this as a book that must be read by every established or aspiring fast bowler and cricket coach.

D. K. LILLEE

The Authors

DARYL FOSTER (MEd).   Daryl has been coach of the very successful Western Australian Sheffield Shield cricket team from 1975 to 1989. This has spanned the era of such great players as Lillee, Marsh, Hughes and Alderman. He has been the most successful coach in Australia, in recent times, in that he has coached Western Australia to eight Sheffield Shield victories over the past fourteen seasons. Western Australia has just won three consecutive Sheffield Shields from 1987 to 1989.

DAVID JOHN (MEd).   David has been the physical educator for the Western Australian Sheffield Shield team from 1985 to 1989. During this very successful period in Western Australian cricket, his training and rehabilitation programmes have become the standard against which other cricket programmes are compared.

BRUCE ELLIOTT (PhD).   Bruce is a senior lecturer in sport biomechanics at The University of Western Australia and the Chairman of the Western Australian Institute of Sport. He is at the forefront of sport biomechanics which examines the most efficient techniques used by sportspeople in their particular event. Bruce has published prolifically in the sports of cricket and tennis.

KEN FITCH (MBBS, MD).   Ken is one of the world's leading sport physicians. He was the Australian Olympic team physician from 1972 to 1984 and is presently a member of the International Olympic Committee Medical Commission. Ken's service to sports medicine has been acknowledged in that he has been made a Fellow of the Australian Sports Medicine Federation and the Australian College of General Practice.

SANDY GORDON (PhD).   Sandy is a lecturer in sport psychology at The University of Western Australia and since 1987 has been the Western Australian Sheffield Shield team mental skills training consultant. He also provides an advisory service to coaches and players of other Western Australian representative teams.

BRIAN BLANKSBY (Ph.D).   Brian is the Head of the Department of Human Movement Studies at The University of Western Australia. He has been involved in sports coaching and research in applied anatomy and sport science, and has published more than a hundred articles in these areas.

Acknowledgements

The authors thank David Treloar and Keith Punch, both very knowledgeable cricket lovers, for their assistance in the preparation of this book, and also the Australian Sports Commission (National Sport Research Program) for funding research into back injuries to fast bowlers in cricket.

CHAPTER 1

The Fast Bowler

BRUCE ELLIOTT

His mighty thighs lifted like the pistons of the Melbourne Express, his boots hammered the ground like a mob of wild horses and when he bowls there's a smell of burnt leather in the air.

(Stivens, 1955)

How often does the rise to cricketing power coincide with the advent of a pair of fast bowlers? Miller and Lindwall, Tyson and Statham, Trueman and Statham, Hall and Griffiths, Proctor and Pollock, Roberts and Holding, Lillee and Thomson, and today not just two but a fearsome foursome of Marshall, Ambrose, Patterson and Walsh from the West Indies have dominated cricket at different times. As Tyson, a prominent fast bowler himself, said, '. . . their fearsome speed makes them legends in their own time'.

Mastery of the proper technique(s), training procedures and psychological make-up are all essential if a bowler is ever to reach full potential. Understanding these factors will better equip and prepare bowlers to learn the art of fast bowling and coaches to detect and correct flaws in the bowling action.

While the rewards of success are easily identified, the stresses and strains placed on the body during fast bowling are a major concern to the bowler. Fast bowling is an impact sport, with the bowler experiencing a series of collisions with the ground in the run-up phase. This reaches a crescendo of two large collisions with the landing of the back-foot and then the pounding of the front-foot during the delivery stride. These two major collisions generate forces of approximately five times the weight of the body. These forces are then transmitted through the bones, cartilages, tendons and muscles via the foot, ankle, knee, hip and back joints. At the same time, the trunk is stretching sideways, bending and twisting in an endeavour to achieve maximum delivery speed.

It is not surprising then that injury may occur, even though each repeated trauma to the body is, on its own, less than the threshold that joints and musculoskeletal tissues can normally tolerate. However, the cumulative effect of many overs each day, over a series of days, and over a number of months throughout a packed local season or over a year at the international level, can gradually create damage just as

1

surely as a hard, single blow. The heavy schedules and large number of games can prevent adequate recuperation and the healing process. The vast majority of sports injuries are those to the musculoskeletal tissues and the most troublesome of these are the 'overuse injuries' such as stress fractures. Indeed, stress fractures are noted in humans, racehorses and greyhounds and in no other animals. The common feature is that they are the only animals who train and compete at maximum effort, despite pain. Thus, in cricket, the enthusiasm of the human spirit, which is so important to a fast bowler, may well be one of the underlying causes of an overuse injury.

Fast bowling can cause trauma to various parts of the body because of overuse but the lower back is particularly vulnerable. The constant sprinting, twisting, stretching and straining that are part of being a fast bowler puts the body through a rigorous test with every delivery.

Overuse injuries which previously were mainly found in adult cricketers are becoming more common place in children and adolescents. Micheli (1983) listed factors which may lead to an overuse injury in either the young or adult fast bowler:

- An abrupt change in training intensity, duration or frequency may precipitate injury.
- The physiological state of the fast bowler may provide a risk factor, whether it be an imbalance of strength or flexibility or some anatomical misalignment.
- The selection of footwear and the type of training/playing surface may be risk factors (e.g. no spikes on slippery surfaces or rough crease areas).
- The existence of an injury or an associated disease will increase the risk of further injury.

While tending to highlight the epidemic proportions that injuries to would-be fast bowlers have reached, it is not all a picture of gloom. The authors of this book firmly believe that fast bowling is no more likely to produce injuries than many other sporting activities if proper technique, appropriate physical preparation and sensible bowling spells are undertaken. Then, if pain is apparent when bowling, appropriate treatment and recuperation practices can be adopted to minimize long-term or recurrent injury problems. **This book outlines how a fast bowler should prepare in order to bowl successfully and to minimize the risk of injury. Because it is impossible to remove all possibility of an injury occurring, appropriate rehabilitation procedures are outlined to restore the bowler back to his/her fastest and best.**

CHAPTER 2

Fitness for Fast Bowling

DARYL FOSTER and DAVID JOHN

Physical fitness is an important aspect in preparing for a career as a fast bowler. Today the fast bowler must **train to play** whereas many fast bowlers of yesteryear were able to **play to train** because they led a more active lifestyle—for example, they may have worked as miners or timber workers, rather than students or office workers. Physical fitness, technique practice and the proper mental preparation are all important components in successful fast bowling. Because fast bowling causes severe pressure and trauma to muscles, joints and bones, specific fitness programmes are essential and should be superimposed upon a strong general level of fitness. This follows the adage of **generalization** before **specialization.**

Poor physical preparation can predispose one to premature tiredness and cause a lack of concentration, resulting in a lesser performance. Lack of fitness can also be a cause of injury. Therefore, it is necessary for both the aspiring and the experienced fast bowler to attain high levels of aerobic and anaerobic fitness, muscular endurance and strength, plus retain adequate levels of flexibility. The well-prepared fast bowler will then be able to operate at a high level of intensity over repeated spells without becoming unduly fatigued. A high level of aerobic fitness will also assist the fast bowler to bowl effectively throughout the season. Training programmes to improve muscular endurance, strength and flexibility enable the muscle groups involved in fast bowling to contract sequentially in a co-ordinated way. This generates the power necessary for successful fast-bowling performance without excessively stressing muscles and joints.

The great fast bowlers have generally been 'fitness fanatics' and have paid special attention to all aspects of their training programmes. Dennis Lillee, in *The Art of Fast Bowling* (1977), states, 'What a vital part of my game physical fitness has been ... Young fast bowlers must realize the onus is on them to start putting in some hard work ... Natural ability will carry you just so far ... It is literally survival of the fittest in the sphere of fast bowling.'

Bob Willis, the great English fast bowler, agreed with Lillee. In his book *Fast Bowling* (1984), Willis noted that 'A fast bowler has to be fitter than anyone else in the team ... There is no hidden secret about the way I keep fit to bowl at the top level ... Basically it involves running, sprinting, strength and stretching exercises.'

The modern-day fast bowler, junior or senior, needs to pay even more attention to

3

fitness than great fast bowlers of an earlier era. As our lifestyle becomes more sedentary, the specific fitness requirements for fast bowling become even more important.
For the young fast bowler in particular, graduated fitness training is of vital importance. Most young fast bowlers commence bowling during the time the body is
maturing rapidly, and this must be considered when developing a programme. A
progressive physical build-up will help to reduce the problems associated with overuse injuries in this developmental phase.

Fast bowlers have always created great interest and have, in the main, been the
match winners in this modern era of cricket. The recent exploits of Lillee, Thomson,
Snow, Willis, Botham, Hadlee, Imran, Kapil Dev and the magnificent string of
West Indian fast bowlers have pleased crowds with their dynamic performances. All
of these great fast bowlers have paid close attention to their individual fitness
requirements. They have realized that their role in the modern game of cricket,
including the one-day fixtures, is very demanding.

Physical fitness training needs to be blended with technique practice in the nets
and match play in order that a sensible, progressive and specific training schedule is
followed by the individual fast bowler at all levels of the game. A carefully kept
training diary is always beneficial. Well-designed programmes should include all the
parameters of physical fitness and be carried out in the off-season, pre-season and
in-season for success in fast bowling.

The physical fitness requirements for fast bowling should include an understanding of the:

1. physiological requirements of fast bowling;
2. physical requirements of fast bowling;
3. physical and physiological assessment of the above factors;
4. principles for developing fitness;
5. training programmes for fast bowlers;
6. pre-game and pre-training considerations;
7. game workload requirements.

1. Physiological Requirements of Fast Bowling

Fast bowling requires several physiological capacities to be developed. These
include:

(a) aerobic capacity;
(b) anaerobic capacity;
(c) muscular strength;
(d) muscular endurance;
(e) flexibility.

(a) AEROBIC CAPACITY

The aerobic (oxygen) system is the major supplier of energy for physical activities
which must be sustained over extended periods. This energy system will be limited,
however, by the rate at which oxygen can be delivered to the working muscles by the
heart and lungs. The amount of oxygen that can be consumed per minute (often

expressed as maximum oxygen uptake, or max. $\dot{V}O_2$) will determine the capacity of one's aerobic (endurance) system.

(b) ANAEROBIC CAPACITY

The anaerobic energy system is the predominant energy source during high-intensity, short-duration exercise such as during short sprints or the run-up when bowling. This system relies primarily upon the utilization of high-energy phosphates stored in the muscles.

(c) MUSCULAR STRENGTH

Muscular strength is defined as the force or tension a muscle or group of muscles can exert in one maximal effort. This may be classified as either relative or absolute strength. Relative strength is important in activities which require high strength-to-weight ratios such as gymnastics and wrestling. Absolute strength is important for weight lifters or shotputters. Ideally a fast bowler requires an adequate level of absolute strength to deliver the cricket ball at high velocities over an extended period.

(d) MUSCULAR ENDURANCE

Muscular endurance is defined as the capacity of a muscle or muscle group to perform repeated contractions against a load or for a single contraction held over an extended period. Given that fast bowling is often a repeated movement over a long time period (ten six-ball overs), muscular endurance is an important component in the physical make-up of a fast bowler. Endurance should be developed in those muscles which make a major contribution to the fast bowling action, particularly the shoulder, lower back, abdominal and lower limb muscles.

(e) FLEXIBILITY

Flexibility is another important component of athletic performance. It may be defined as the range of motion about a joint, such as the ability to touch one's toes, or the range of movement about a number of joints required to perform a skill. Fast bowlers should develop optimal levels of flexibility in the shoulder, back and hip joints, and the surrounding muscles to enhance performance and reduce the possibility of injury.

2. Physical Requirements of Fast Bowling

Some physical components required by fast bowlers can be improved through specific training programmes but others are inherited at birth and may be more difficult or even impossible to change. Physical characteristics which are important to fast bowlers include:

(a) optimal levels of body fat;
(b) good posture;
(c) superior co-ordination.

(a) BODY FAT

Both males and females have individual but fairly constant patterns of fat deposition. Some body fat is necessary as an energy source, to protect the body from injury and to assist in heat retention. Too much body fat, however, may be detrimental to performance in two ways:

• excess fat does not contribute to energy production;
• the body must consume energy to support excess fat and carry it about.

Fast bowlers should therefore be relatively lean and carry only minimal levels of body fat.

(b) POSTURE

Posture involves the anatomical relationship between the skeletal and muscular systems when at work or at rest. A fast bowler with well aligned posture demonstrates a state of skeletal and muscular balance which protects the body against injury or progressive deformity when exercising or resting. Conversely, poor posture involves a faulty relationship between various parts of the body which produces increased stress and strain on supporting structures. For example, fast bowling is a unilateral (one-sided) activity which may result in increased development on one side of the spine. Therefore, it may be necessary to perform compensatory exercises to maintain a correct postural balance.

(c) CO-ORDINATION

The co-operative interaction between the nervous system and the skeletal muscles plays an important role in the performance of a fast bowler. The aim is to develop smoothly sequential, neuro-muscular actions of the body segments to produce the powerful forces and explosive movements required for bowling fast. Generally, the necessary interplay and action of the important muscles are a trainable process which can improve with appropriate practice. Remember, practice does not make perfect, but perfect practice does make perfect.

3. Physical and Physiological Assessment

Physiological and physical characteristics should be assessed to determine the relative strengths and weaknesses of a particular bowler. Fitness assessments may be conducted in a laboratory using expensive, sophisticated equipment, or through field tests conducted at the cricket ground. Both forms of fitness testing can provide useful information to the player and coach. For example, the coach receives feedback regarding the effectiveness of training programmes while the player can be motivated to improve any weaknesses.

There are numerous laboratory and field tests (Table 2.1) to evaluate physical and physiological characteristics considered to be important in fast bowling. They include:

4. Principles for Developing Fitness

After a physical and/or physiological profile has been identified through testing, a specific exercise prescription is then developed to rectify weaknesses while maintaining strengths. A properly designed programme should employ five basic principles of training which will ensure that the bowler receives maximum benefit.

TABLE 2.1

PHYSICAL AND PHYSIOLOGICAL ASSESSMENT — LABORATORY AND FIELD TESTS

Measure	Lab. Test	Field Test	Field Norms*		
Aerobic capacity	Max. $\dot{V}O_2$ run on treadmill PWC_{170} bike test	12 min. run	V.G. > 3600 m G. > 3300 m Av. > 2700 m P. < 2400 m		
Anaerobic capacity	Exertech 10 s, 30 s bike test	Phosphate test (see Dawson *et al.*, 1984, for typical test)	V.G. > 90 s G. > 75 s Av. > 60 s P. < 45 s		
Muscular strength	Biodex, Kincom one effort	Maximum lift in one effort			
Muscular endurance	Biodex, Kincom repeated efforts	Number of sit-ups (S-U) or push-ups (P-U) in 60 s		**S-U** V.G. > 55 G. > 45 Av. > 35 P. < 25	**P-U** > 50 > 43 > 32 <25
Muscle power	Biodex, Kincom Margaria-Kalamen stair test	20 m sprint 40 m sprint		**20 m** V.G. <3.0 s G. 3.2 s Av. 3.4 s P. > 3.6 s	**40 m** <5.6 s 5.8 s 6.0 s >6.2 s
Flexibility	Goniometry measures	Sit and reach test	V.G. + 16 cm** G. + 8 cm Av. + 2 cm P. < − 2 cm		
Posture	New York posture rating	Subjective assessment			
Body fat	Underwater weighing	Skinfold measures using calipers (triceps, subscapular, suprailiac and mid-abdominal)	**Sum (four Sites)** V.G. <35 mm G. 45 mm Av. 55 mm P. >70 mm		
Co-ordination/agility	Dynamic balance platform	Illinois agility run	V.G. <15.0 s G. 15.4 s Av. 16.0 s P. >16.4 s		

* These are adult norms: V.G. (very good); G. (good); Av. (average); P. (poor).
** Relative to the toes.

(a) SPECIFICITY

A training programme must be specific for improving or maintaining the energy system or systems predominantly used during performance of the sport activity in question. Given that fast bowling is predominantly an anaerobic activity which requires an aerobic base, training programmes for fast bowlers should develop both energy systems. Aerobic fitness should be improved during the pre-season and maintained through the season, while anaerobic fitness should be improved during the season.

(b) FREQUENCY

The more frequent the training programme, the greater will be the fitness improvements. However, it is recommended that when training for endurance activities, one should only train on three or four days per week, or alternate days. When one seeks to improve the anaerobic system, three days per week are sufficient. Training more frequently than this (four days per week) does not enable the body to recover completely in between exercise bouts.

(c) INTENSITY

The intensity of training is directly related to improvement in overall fitness. Measurement of training intensity is best gauged by monitoring heart rates. It has been determined that, to gain training benefits, exercise during the training programme should be intensive enough to cause the heart rate to reach 65–90% of the maximum heart rate.

(d) PROGRESSIVE OVERLOAD

Exercise intensity and/or resistance must be progressively increased throughout a training programme to a degree which is commensurate with the player's fitness improvement. The player's strength and endurance will only increase when training occurs against workloads that are above those normally encountered.

(e) VARIETY

Variety has an important role in the training programme in maintaining and promoting interest. A cricket season, including the off-season and pre-season, could conceivably last for eight to nine months and, without a varied training programme, staleness and complacency may develop. Therefore, variety should be added to alleviate these problems, in the following ways:

- Change the training venue and training times.
- Introduce novel and challenging cricket activities.
- Incorporate games into training.
- Place more responsibility on players to develop training methods.
- Use different methods to develop fitness such as aerobics, swimming, cycling.

5. Training Programmes for Fast Bowlers

Various programmes to develop and maintain specific fitness components required for fast bowling are provided below. These programmes are suitable for all fast bowlers but should be modified after considering age, training background, body build, fitness level and fitness requirements. Any adjustments to programmes should only be made after consultation with suitably qualified persons such as a physical educator or coach.

(a) RUNNING PROGRAMMES

Running programmes can improve the aerobic and anaerobic energy systems when using appropriate training principles.

Aerobic training. This involves continuous running for relatively long distances to develop endurance. A fast bowler's goal should be to run 10 km at an intensity fast enough to maintain a heart rate of 70–85% of maximum heart rate (probably about 4–5 min./km pace).

To develop endurance capacity rapidly, the fast bowler should run seven days per fortnight with a rest day between each run. Then, either the distance or the speed of the run should be increased as endurance improves. The bowler may start this build-up by jogging/walking for a total of 30 minutes. A gradual increase in distance covered and speed should enable the goal to be reached in approximately two to three months. The establishment of an aerobic base for performance is therefore ideally carried out during the off-season and early pre-season, and maintained by at least one run per week during the season.

Fartlek training. This involves alternating between fast and slow running speeds which can develop all of the energy systems. Fartlek running does not require manipulation of work or rest periods. Instead, the bowler alternates speed based on how he or she feels at the time. For example, one runs in a natural surrounding such as a park or beach and enjoys the view. When feeling comfortable one puts in a burst of 10–30 s and then eases back until feeling comfortable again before repeating the effort. This form of running is ideal for general conditioning in the early pre-season and to provide variety in training.

Interval training. This involves pre-determined spacing of exercise and rest periods to develop the aerobic and/or anaerobic energy systems. An interval training programme can be modified in terms of:

- distance to be covered in each repetition;
- interval of departure: length and type of rest period in between efforts;
- repetitions: a pre-determined number of repetitions;
- time and intensity of the work period.

Briefly, an interval training system to meet the requirements of training for fast bowling would be as shown in Table 2.2.

Longer, slower intervals are suitable during the off-season and early pre-season (1500–400 m), while shorter, faster intervals are preferred during the season (100–200 m).

Sprint training. This is used to develop anaerobic power and capacity and muscular strength as the athlete performs repeated sprints at maximal or near maximal speed. Two different forms of sprint training are appropriate for fast bowlers:

- Interval sprints—a method of training whereby a bowler alternately sprints and then jogs over a given distance, e.g. 4 x 50 m sprints, 50-150 m walk/jog between sprints.

- Acceleration sprints—involve a gradual increase in running speed from jogging (25-30 m), to striding (25-30 m) to sprinting (25-30 m, plus an easing down over 35-50 m). A complete recovery, while walking slowly, follows each series until breathing is comfortable.

Sprint training is usually performed during the season to increase speed and muscle strength and to maintain anaerobic fitness.

TABLE 2.2

INTERVAL TRAINING PROGRAMME

Anaerobic

Intensity of exercise	1.5–3 s are added to the bowler's best time for 20–100 m (about 85% of best time)
Rest period	Work to rest ratio 1:3 e.g. run 15 s; walk or rest 45 s
Repetitions	High number of repetitions (8–10)
Sets	Multiple sets (2–3), 3–5 min. rest between each set

Aerobic

Intensity of exercise	10–15 s slower than average 400 m time for each repetition (work at about 80% of maximum speed) e.g. For one who runs 400 m in 60 s, 80% = (60 + 60/5)s = 72 s
Rest period	Work to rest ratio 1:1.5 or 1:2 e.g. run 72 s; jog/walk for 2 mins
Repetitions	Low number of repetitions (4–8)
Sets	1–2 sets, 5 min. rest between each set

(b) FLEXIBILITY PROGRAMME

Flexibility varies markedly between individuals and, in fact, may vary significantly in the one person. That is, a fast bowler may have excellent shoulder flexibility but very poor hip flexibility. Specific stretching exercises will improve joint and muscle flexibility if performed regularly. How to stretch:

- Stretch slowly and with control.
- Never bounce.
- Stretch to a point of tension, never to the point of pain.
- Breathe in a normal, rhythmical manner.
- Don't compare your flexibility with others.
- Hold each stretch for 10–30 s.

There are two phases of stretching which should be understood before beginning the stretching programme:

- **Primary phase.** Stretch to the point where one feels a slight tension in the muscle. Hold this for 10–30 s so that the tension decreases, or ease off until feeling comfortable. This reduces muscular tightness.
- **Secondary phase.** After the primary phase, move further into the stretch until there is tension in the muscle again. Hold for 10–30 s. This increases flexibility and prepares the muscle for exercise.

Good flexibility is important when performing any exercise because it reduces susceptibility to muscle strains and tears by warming and readying the muscles for activity. Therefore, a regimented stretching programme should be undertaken prior to any bowling.

A general stretching programme, appropriate for fast bowlers, is outlined as follows:

Calf stretch. Stand with one foot in front of the other and apart; front knee bent slightly and the rear leg straight. Slowly move the hips forward until a stretch is felt in the calf muscles of the back leg. Be sure to keep the rear heel on the ground and the toes pointed straight ahead. Hold the stretch for 10–30 s and relax. Repeat for other limb.

Achilles stretch. Assume the same position as for calf stretch except that the back knee is slightly bent. Hold for 10–30 s and relax. Repeat for the other limb.

Quadricep and knee stretch. Standing upright, place the foot in the hand from the same side of body (i.e. hold top of right foot with right hand). Kick downwards with the foot and resist any movement with hand. Hold for 10–20 s and relax. Repeat for other limb.

Hamstring stretch. Sitting on the ground with one limb straight (toes pointing upwards) and other bent, flex at the hips and bring the forehead towards the straight knee until feeling a mild stretch in the hamstrings. Hold the stretch for 10–30 s. Repeat for other limb.

Adductor (groin) stretch. Put soles of feet together with heels a comfortable distance from groin. Put hands around feet and lean forward until a stretch is felt in the groin. Hold for 10–30 s and relax.

Gluteal stretch. Assume a back lying position with both legs straight. Pull the right knee to the chest keeping the head on the floor. Hold for 10–30 s and relax. Repeat using other limb.

Lumbar stretch. Assume a back lying position with both knees flexed to 90°. Drop both knees down to the right side of the body, keeping both shoulders in touch with the ground, and hold for 10–30 s. Repeat stretch to other side.

Lower back stretch. Assume starting position for lumbar stretch. Place the right ankle over left knee and use the right leg to pull left leg toward floor until feeling a stretch along the side of the hip and lower back. Keep both shoulders in touch with the floor and hold for 10–30 s. Repeat to other side.

Upper back stretch. Assume starting position for lumbar stretch. Pull both knees toward chest and rock back and forth on the upper back ten times.

Shoulder stretch:

- UPWARD—in standing position, interlace fingers above the head. With palms facing upward, push upwards and slightly backwards. Hold for 10–30 s and relax.

- POSTERIOR—in standing position, interlace fingers behind the back. Slowly turn elbows inward while straightening arms upward. Hold for 10–30 s and relax.

- TRICEP—in standing position, place arms above head and bend the right elbow 90°. Pull the right elbow downwards, resisting this pull with the left hand. Hold for 10–30 s and relax. Repeat in opposite direction.

Side stretch. With arms overhead, bend from the hips to side. Hold for 10–30 s and repeat for the other side.

(c) STRENGTH PROGRAMME

Muscular strength will only improve if muscles are exercised at close to the maximum force they are capable of generating. Muscle overload can be applied with different apparatus such as free weights, pulleys, springs or isokinetic devices, but it is the intensity of the overload which generates strength improvements rather than the particular method. Two of the more popular forms of strength training for fast bowlers are as follows.

Circuit Training

This involves lifting light weights for a set period of time, where emphasis is placed on providing a general conditioning programme to improve muscular endurance, body composition, some strength and cardiovascular fitness rather than absolute strength improvement. Typically, circuit training involves lifting light weights (40–50% of maximum) as many times as possible in a set time (30 s), followed by a brief rest (15 s), and then one moves to the next exercise station. Generally, eight to twelve different exercises are available and the circuit is repeated two to three times. When no weight equipment is available the individual may use his/her own body weight (e.g. sit-ups, push-ups, dips). Often, circuit training should precede any weight lifting because:

- lighter weights are used;
- it enables the bowler to learn correct lifting techniques;
- it more easily allows limbs to move through a fuller range of motion than would be possible with heavier weights;
- it enhances general fitness prior to the development of the specific requirements of fast bowling.

A circuit requiring no weights but which is appropriate for fast bowlers is outlined as follows.

Commence by completing two sets of the following exercises and increase this to three sets after three to four weeks. Increase the training load by completing more repetitions in each set.

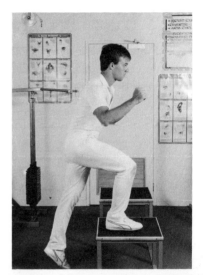

Step-ups: Step up and down thirty times in 30 s and recover for 30 s. Repeat using other leg to step.

Dips: Dip up and down five to fifteen times in 20 s and recover for 30 s.

Sit-ups: Sit up and back thirty times in 1 minute and recover for 30 s.

Leg-overs: Complete fifteen leg-overs in 30 s and recover for 30 s. Repeat to the other side.

Reverse curls: Complete fifteen curls in 30 s and recover for 30 s.

Alternate leg lifts: Complete ten lifts of each leg in 30 s and recover for 30 s.

Push-ups: Press up and down ten to fifteen times in 20 s and recover for 30 s.

Crunches: Complete fifteen crunches in 30 s and recover for 30 s.

Alternate knee lifts: Complete ten lifts of each leg in 30 s and recover for 30 s.

Bench blasts: Complete ten to fifteen blasts in 20 s and recover for 30 s. Repeat, leading with other leg.

Weight Lifting

This involves the use of barbells, dumbells and/or weight machines to strengthen specific muscles. Relatively light weights should be used during the early stage of a strength programme to reduce the possibility of muscle injury. Generally, a weight that will allow twelve to fifteen repetitions of an exercise should be used at the commencement of a weight programme for two to four weeks. When the muscles have adapted and lifting techniques are learned, the number of repetitions can be progressively reduced to six to eight as more weight is added. A sample weight training programme for fast bowlers is outlined in Table 2.3 with appropriate weights and numbers of repetitions to be completed for those starting out and those with more lifting experience.

- Select a weight with which fifteen repetitions of the first set can be completed but the final three to four repetitions should be difficult to complete. During sets 2 and 3, a lesser number of repetitions may be possible. When three sets of fifteen repetitions can be completed it is time to add more weight.
- Begin by doing two sets and build up to three sets after two to four weeks of Level 1 and progress slowly to Levels 2 and 3.
- Two to three times per week.
- Work to rest ratio is 1:3.
- Warming up and cooling down are very important.
- Begin and end the programme with a flexibility programme.

TABLE 2.3
PROGRAMME BASED ON WEIGHTS

	Beginner Levels			Advanced Levels		
	1	2	3	1	2	3
Bicep curls	15	12	10	12	10	8
Upright rowing	15	12	10	12	10	8
Dumbell flys	15	12	10	12	10	8
Inclined sit-ups	10	15	20	12	18	25
Tricep extensions	15	12	10	12	10	8
Lateral raises	15	12	10	12	10	8
Twisting sit-ups (inclined)	10	15	20	12	18	25
Leg curls	15	12	10	12	10	8
Leg extensions	15	12	10	12	10	8
Shoulder press	15	12	10	12	10	8
Back extensions	10	15	20	10	15	20
Bench press	15	12	10	12	10	8
Lat. pull down	15	12	10	12	10	8
Crunches	10	15	20	12	18	25
Pulley work*	15	15	15	15	15	15

Bicep curls

* Adopt the normal fast bowling stance (back foot parallel, shoulders side-on) and practise the delivery action with the pulleys. It is important to work both the bowling arm and the non-bowling arm.

Upright rowing

Dumbell flys

Inclined sit-ups

Tricep extensions

Lateral raises

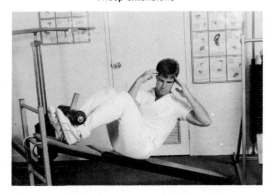

Twisting sit-ups (inclined)

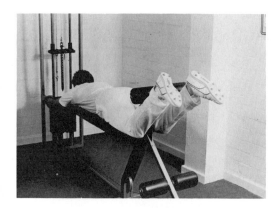

Leg curls

Leg extensions

Shoulder press

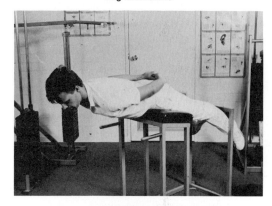

Back extensions

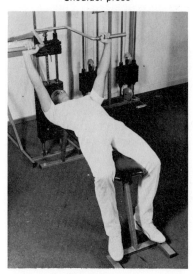

Bench press

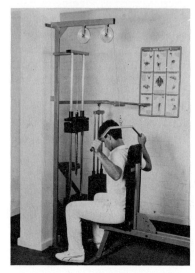

Lat. pull down

Crunches

Pulley work

6. Pre-game and Pre-training Considerations

Five general areas should be considered prior to training or a game.

(a) warm-up;
(b) fluid intake;
(c) food;
(d) general health;
(e) equipment.

(a) WARM-UP

It is important to warm up adequately prior to commencing any bowling or strenuous exercise. The muscle and body temperature should be increased through mild exercise such as jogging or calisthenics until a light sweat is achieved. This serves to facilitate blood flow and increase the availability of oxygen (nutrients) to the working muscles. Stretching exercises which are specific to fast bowling requirements should then be performed. Examples of these exercises can be found earlier in this chapter. The warm-up is then completed by practising simple skills such as rehearsing the delivery stride of the bowling action four to five times.

(b) FLUID INTAKE

Cricket is a summer sport and is played often in hot and/or humid conditions. Dehydration and heat stress may occur if precautions are not taken. On a hot and/or humid day a bowler should:

• be adequately hydrated prior to exercise (drink 500 ml of water);

• always drink at the scheduled drink breaks even if not thirsty (300–500 ml). More cold water should be consumed at breaks in play;

- replace all fluid lost over a day's play primarily with water rather than soft drink or alcohol at the end of the day's play;
- provide protection from the sun and limit fluid loss by wearing suitable clothing which includes a hat and a long-sleeved shirt.

(c) FOOD

Eating prior to exercise may either enhance or limit performance depending on the type of food eaten and the time this food was consumed with respect to the start of play. Some general rules for eating are:

- The pre-game meal should be eaten at least 2.5 hours prior to play.
- This meal should consist mainly of carbohydrates.
- No sugar (cakes, chocolate bars, soft drinks) should be ingested in the hour preceding play.
- Liquid pre-game meals and/or during-game meals (e.g. Sustagen) should be considered as part of a bowler's choice of foods.

(d) GENERAL HEALTH

A good state of general health is necessary if a bowler is to perform well and there are several conditions under which a player should not participate:

- Illness—a bowler should not train when ill.
- Injury—a bowler should not perform any activities that would aggravate an injury.
- Alcohol—consumption of alcohol prior to or during exercise has a deleterious effect on reaction time, cardiovascular endurance, speed, power and co-ordination.
- Smoking—smoking prior to or during a game may limit performance by restricting the breathing airways and decreasing the amount of oxygen that can be carried by the blood.

(e) EQUIPMENT

Boots and socks are among the most important equipment for a fast bowler. Inadequate foot protection may lead to blisters, bruises, shin splints or even more serious problems. A fast bowler should:

- wear well-fitting, comfortable boots;
- 'break in' a new pair of boots at training rather than in a game;
- wear two pairs of socks to provide added protection;
- always treat blisters or other foot problems immediately they occur;
- replace worn sprigs on boots.

7. Game Workload Requirements

The fast bowler is a scarce and valuable resource to any team and should be treated with care by the captain. It is true that spinners rub the skin off their fingers, batsmen can be felled by bouncers, and fieldsmen might be injured taking catches, but fast bowlers are far more prone to injury. Scarcely any single catastrophe can jeopardize a match as much as a fast bowler being unable to continue to bowl due to injury.

The likelihood of injury can be increased by fatigue, by overwork, by trying to bowl too fast or by dangerous playing surfaces, so the captain must endeavour to control these factors.

(a) LENGTH OF BOWLING SPELL

When a fast bowler opens the attack, both the bowler and the captain should share the same understanding as to the number of overs that will make up the spell. The objective of obtaining the initial breakthrough is the first 'performance goal' for an opening bowler. Naturally, unless the bowling is well below standard or a complete reconsideration of tactics is called for, the spell for anyone over eighteen years will generally extend up to approximately eight overs (for more complete bowling spell guidelines, see Chapter 6).

Towards the eighth over of a spell, the captain and the bowler should talk about any tiredness or difficulty the bowler is experiencing in maintaining top performance. The bowler should not hesitate to ask for a break because there is no advantage in the mock heroics of soldiering on if the beginnings of pain or stress can be sensed and there is another bowler available to come into the attack. A fast bowler should also be given adequate warning of the likelihood of being reintroduced into the attack so that sufficient time is available for a warm-up to be carried out.

The captain needs a close understanding of fast bowlers, so as to know when to call for the extra effort without undue risk of injury or exhaustion from which it will take too long to recover. Similarly, bowlers need to understand themselves in order to know when not to give up too readily but keep plugging on through barriers of discomfort.

Early signs of fatigue may be a wandering line or length, or a slowing of pace. The captain, in order to see this and also to be able to judge other aspects of the quality of the bowling, should be fielding in an appropriate position such as the slips or mid-off. Otherwise, it is common to rely on the wicketkeeper for information.

When noticed early, these signs of fatigue often can be overcome by encouragement or firm advice. The body is the servant of the mind when fatigue is not severe, and a bowler whose captain has made the right comment at this stage often can rise to the challenge and carry on with renewed energy and control.

In the sometime extreme Australian conditions, medium-fast bowlers have been known to bowl fifteen or more overs unchanged, even in club cricket where physical preparation is less than perfect. This can mean bowling unchanged throughout a whole session till tea. Genuinely fast bowlers operating at full speed should not be asked to do so much. This is especially so if they are in their teens, because the risks of injury make it foolhardy to do so. Also, a genuine fast bowler needs to retain a hostile breakthrough mentality and not be reduced to a stock bowler unless, as happens from time to time, it is necessary to carry the extra burden.

The break from fast bowling, when it comes, is recommended to be of about an hour (see Chapter 6 for bowling guidelines).

(b) HOW FAST TO BOWL?

Although a potential speedster is unlikely to develop into one if not given the chance to whizz them down, it is a tactical mistake for a captain to allow a young fast bowler to bowl as fast as possible for as long as possible. From the team's point of view, sheer speed may be less effective than controlled line and length in getting wickets, and many bowlers lose their out-swinger if they try to bowl flat out, due to the loss of control over precise body positions at the moment of delivery. From the bowler's point of view, aiming for sheer speed increases the risk of strain and fatigue. On the other hand, few bowlers have the ability to generate genuine speed and, therefore, those who do should be given every incentive to use it. Later in their careers they may need to settle for line and length.

CHAPTER 3

Fast Bowling Technique

BRUCE ELLIOTT and DARYL FOSTER

Fast bowling success is greatly influenced by the way the ball is delivered. Perhaps the main objective of the teacher or cricket coach should be to develop a 'good technique' in young fast bowlers. Thus, a thorough understanding of all the components of 'good technique' in the bowling action is necessary. Then, the underlining **personal characteristics and flair of the young bowler, together with the proper mechanics of fast bowling, can be integrated to develop techniques which suit that particular individual.** In this chapter, the bowling action is divided into a number of different sections to dissect the mechanical features of fast bowling. Then in Chapter 4, the basis of swing bowling is considered.

1. The Grip

The basic grip for fast bowling is to have the index and middle fingers on the seam and to place the thumb underneath the ball (Figure 3.1). The fourth and fifth fingers act to balance the ball in the hand in a relaxed fashion. The significance of the position of the seam and its influence on the swing of the ball are dealt with in the next chapter.

Figure 3.1: The basic grip

2. The Run-up to Back-foot Landing

The aim is to have an approach-run usually between 15 m and 30 m which builds up to an optimum speed about three to four strides before delivery. During the run-up the eyes are fixed on the base of the batsman's off-stump. A bowler must then learn to adjust where the ball is to be pitched in order to bowl both good line and length. This will require practising in the nets, with the bowler focusing on the base of the off-stump while striving to hit a marker placed on the pitch at good length for a particular line. The bowler must practise these tasks from a full run for at least part of a session so that correct front-foot placement can be practised to avoid bowling no-balls.

Davis and Blanksby have shown that the fastest bowlers tend to have a faster approach velocity than medium-fast bowlers and therefore this aspect of fast bowling must be practised. Lillee has warned bowlers that a run-up which is too fast makes it difficult to attain a side-on delivery action, whereas too slow a run-up reduces the contribution of the run-up speed to the ball. Bowlers should also be aware that they slow down considerably during the second last stride in order to achieve this side-on position. Lillee was measured moving into his delivery stride at a speed of 5 m/s (17.6 km/h) after reaching a maximal velocity of 9 m/s (32 km/h, a speed of approximately 80% of maximum). During the delivery leap a cross-over step is performed. The body should begin to assume a side-on position while in flight so that at back-foot impact the shoulders and hips are already pointing down the pitch. Emphasis needs to be placed on:

- rhythm;
- a gradual build-up of speed with the maximum approach-speed recorded three to four steps from the delivery stride. Bowlers who try to increase run-up speed at back-foot landing will have difficulty in attaining a side-on position;
- a final run-up speed of 4–5 m/s at release;
- focusing on the base of the off-stump.

At back-foot impact, the back-foot should be almost parallel to the rear crease (Figure 3.2A) if a side-on position (body facing the stumps at the bowler's end) is to be achieved. A back-foot placement pointing down the pitch or towards square leg (right-handed batsman) will create a very open-chested action (Figures 3.2B and 3.4A). **Bradman (1958) summed up this phase of the delivery when he stated that the wind-up is the most important phase in attaining a side-on bowling action because errors at this stage cannot be compensated for later in the movement.** The emphasis should be on:

Side-on Technique:
- a back-foot angle parallel with the rear crease;
- having a line through the shoulders and hips pointing down the pitch.

Front-on Technique:
- a back-foot angle pointing approximately down the pitch;
- having a line through the shoulders and hips pointing across the pitch.

Figure 3.2: Back-foot alignment at back-foot impact

A: Side-on technique   B: Front-on technique

3. The Delivery Stride

To date, most coaching manuals give a major emphasis to achieving a side-on position at delivery. It would appear, however, that there are three fast-bowling techniques which are prevalent in cricket.

- The first technique is where a side-on position is attained between back-foot
impact and front-foot impact and is epitomized by Terry Alderman.

- Secondly, there is the front-on bowler who does not vary the open position during
the delivery stride at all. The highly successful Malcolm Marshall uses such a technique and defies many coaching pundits by being able to swing the ball away from
a right-handed batsman.

Both of these styles are acceptable, but the third technique, a combination of the
two, appears to heighten the incidence of injury.

- **In this mixed technique, during the delivery stride, the lower body is front-on but
the shoulders attempt to adopt a side-on position. This increases the load on the
lumbar spine because the body is in a twisted, awkward posture.**

The discussion that follows outlines the actions involved in the two acceptable
techniques and also presents the difficulties associated with the 'not recommended'
mixed action. While it is not always wise to discuss improper actions, because one
should highlight positive attributes, there are so many young bowlers using this
incorrect technique, it is believed that such a discussion is necessary in this instance.

(a) SHOULDER ALIGNMENT
An 'open' (front-on) action reduces shoulder swing and therefore the contribution

of body rotation to final ball velocity. High-speed film of 'side-on' fast bowlers such as Terry Alderman reveals that they rotate their shoulders through arcs of approximately 105° from the landing of the back-foot (Figure 3.5) until release (Figure 3.8). On the other hand, front-on bowlers such as Geoff Lawson rotate only through 90°, and a group of predominantly front-on 'A-grade' fast bowlers were found to rotate the shoulders through 68°. The reduced arc of the shoulder turn reduces the distance over which ball speed can be generated. However, the front-on bowlers compensated for the loss of speed generation in this way by maintaining a higher proportion of the run-up speed at delivery. For the side-on bowler, shoulder rotation is of great assistance in the generation of ball speed. However, a bowler using this action must lose more run-up speed than the front-on bowler due to the 'braking action which must occur immediately prior to back-foot impact' so that a side-on foot and shoulder positioning can be achieved.

A feature common to the mixed technique bowlers that may lead to injury is the change in shoulder alignment from the landing of the back-foot to the planting of the front-foot. Many front-on bowlers dramatically change their shoulder alignment from an open-chested position at the landing of the back-foot (Figure 3.3D) to a less open position (Figure 3.3E) prior to beginning the rotation of the shoulders toward the batsman at the landing of the front-foot (Figure 3.3F). This twisting of the lower lumbar region of the back, particularly when in a hyperextended position (Figure 3.3F), places stress on the lower lumbar vertebrae which has the potential to cause injury.

Furthermore, this partial closure of the shoulders is an admission that the bowler realizes the poor alignment of the initial position and tries to remedy the earlier technique flaw. The fault was probably caused by incorrect placement of the back-foot at back-foot impact. While side-on bowlers may also rotate their shoulders away from the batsman between back-foot impact and front-foot impact, the re-alignment is not of the same magnitude as that recorded for front-on bowlers. Fast bowlers who use the side-on technique should try to emphasize:

- **keeping the shoulders aligned with the pitch such that the front shoulder and hip are pointed at the batsman for as long as possible, thereby remaining side-on during the delivery stride.**

If young fast bowlers use a front-on action they must follow the example set by those West Indian fast bowlers who adopt a front-on shoulder alignment at back-foot impact (Figure 3.4A). They do not attempt to look at the batsman over the non-bowling arm and thus do not hyperextend their spine. Neither do they rotate the torso to any large degree into a more side-on body orientation between back-foot and front-foot impact, which is what places stress on the lumbar region of the back (Figures 3.3E and 3.3F). Bowlers who combine the actions (i.e. who have a front-on foot alignment at back-foot impact while trying to look at the batsman from outside the front-arm in an endeavour to develop a side-on shoulder orientation between back-foot impact and front-foot impact) place the lumbar region under additional stress than is derived from either a purely side-on or front-on technique (Figures 3.3D to 3.3F).

Figure 3.3: Shoulder alignment from back-foot to front-foot impact

Side-on bowling action

A: Back-foot impact   **B:** Mid position   **C:** Front-foot impact

Front-on bowling action

D: Back-foot impact   **E:** Mid position   **F:** Front-foot impact

(b) FOOT ALIGNMENT

On landing, the front-foot should point down the wicket towards the batsman. A line drawn through the toe of the front-foot and the back-foot should be almost a straight line with the middle stump of the wicket at the opposite end (Figure 3.4B).

Figure 3.4: Front-on bowling action—body positions at back-foot impact, front-foot impact and release (Malcolm Marshall)

A: Back-foot impact   B: Front-foot impact   C: Release

Elite fast bowlers have a front-foot alignment slightly to the on-side of the wicket compared to the back-foot. However, 'A-grade' bowlers who tend to have front-on bowling actions have an alignment where the front-foot is displaced to the off-side when compared to the back-foot. Emphasis needs to be placed on:

- **forming a straight line between the front-foot, the back-foot and the stumps at the batsman's end of the wicket.**

(c) STRIDE LENGTH

A delivery stride length of approximately 75–85% of a bowler's standing height seems to be a good rule of thumb. Stride length is most influenced by approach velocity. Those bowlers who approach the bowling crease with an excessive velocity will often have a shorter delivery stride which may inhibit the ability to master a side-on delivery. Emphasis should be placed on:

- **a rhythmic approach that enables a delivery stride length of 75–85% of standing height.**

(d) FRONT-ELBOW MOVEMENT

The non-bowling arm should be almost vertical and placed such that the bowler can look over the outside of the arm at the batsman prior to front-foot impact for a side-on technique (Figure 3.3A) and inside this arm for a front-on technique (Figure 3.4A). At front-foot impact the elbow of the front-arm must be accelerated into the side to assist the rotation of the bowling limb (Figures 3.5 to 3.8). This limb then continues to rotate backwards as part of the follow-through (Figures 3.8 and 3.9). Bowlers who use a side-on delivery action are better able to accelerate the front-elbow into the side than are front-on bowlers, who record lesser velocities for the elbow. Bowlers should emphasize:

- **the side-on position with the trunk flexing laterally and the front-shoulder and limb pointing at the batsman to allow a more forcible use of the non-bowling limb to assist in developing bowling power.**

(e) DELIVERY STRIDE SEQUENCING UNTIL BALL RELEASE

Another key to fast bowling is rhythm. The bowler must add together the movements of different body parts in a sequence that produces the fastest delivery that can be bowled with control.

First, an appropriate run-up speed is needed.

Second, at back-foot impact, the trunk is inclined slightly backwards to allow the front-leg and arm to be in a position to drive vigorously downwards (wind-up phase: Figures 3.2A, 3.2B and 3.4A), particularly when using the side-on action. A more vertical trunk at this point in the delivery stride is common with the front-on action.

Third, between back-foot impact and front-foot impact, the front-leg and arm are thrust downwards, and the trunk and bowling arm begin to rotate toward the batsman. The trunk and bowling arm then continue to rotate after the front-foot is planted (Figures 3.6 to 3.8). The front-arm and bowling arm should rotate together if maximum ball velocity is to be achieved.

The last segment to add to the velocity of the ball is that of the hand. Movement at the wrist, although it does play a role in generation of ball speed, is not as pronounced as is sometimes recommended in coaching manuals. A slightly cocked position at front-foot contact (Figure 3.7) only changes minimally until the hand and the arm are aligned at ball release (Figure 3.8). Emphasis is on:

- **a co-ordinated run-up followed by a rhythmic sequencing of body movements. Despite the appearance of being relaxed, the movements during delivery should be explosive.**

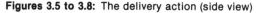

Figures 3.5 to 3.8: The delivery action (side view)

3.5   3.6

3.7   3.8

(f) BODY POSITION AT DELIVERY

At release, a variety of front-knee angles have been recorded for élite fast bowlers. Jeff Thomson and Geoff Lawson, who both demonstrated an almost fully extended front limb (173°) at front-foot contact, were in full extension at the knee by the time the ball was released (180°). Others, however, had a flexed (bent) knee at front-foot contact (150°) and maintained a relatively stable knee angle until the ball was released. Dennis Lillee, Terry Alderman and Malcolm Marshall, to name just a few fast bowlers, flexed the front-knee slightly from angles of approximately 168° at front-foot impact to 159° at ball release. It seems desirable that some knee flexion should occur following front-foot plant to assist in the absorption of the force experienced when the foot impacts the ground (Figures 3.7 and 3.8).

A line drawn through the shoulders at release for high-performance bowlers should be between approximately 10° and 30° past a line parallel with the stumps. As the trunk leans forward and the bowling arm is almost vertical at release, the angle between the trunk and arm should be approximately 200° (Figure 3.8). The wrist joint should change from an angle of approximately 190° during the delivery action to a position where the hand is almost aligned with the forearm at release (180°—Figure 3.8). The wrist is therefore the last segment to play a role, albeit a minor one, in the development of ball speed. At ball release the index and middle fingers should be the last to impart force on the ball. The wrist should be square to the batsman at delivery and should continue to flex after release as part of the follow-through.

(g) THE FOLLOW-THROUGH

It is imperative that the bowling arm follow-through down to the outside of the left leg (for a right-handed bowler) so that it almost brushes the ground (Figure 3.9). The follow-through allows a gradual reduction in body momentum by gradual slowing down of body segments to lessen the stress placed on the joints. During this time the bowler moves off the pitch and recovers balance. At the completion of the follow-

Figure 3.9: Early follow-through (front view)

through the left-hand side of the body (right-handed bowler) rather than the back of the bowler should be visible from behind the bowler. This is the best indicator of a complete follow-through.

4. Captaining the Fast Bowler

Having developed a bowling technique commensurate with body build, it is necessary that the bowler be handled by the captain in a way that elicits optimal performance with a minimal risk of injury.

A number of factors should be considered by the captain when controlling fast bowlers. The length of spell is discussed elsewhere in the book (Chapters 2 and 6) but the following factors are also important:

(a) the choice of bowling end;
(b) where the fast bowler should field;
(c) setting the field;
(d) where the fast bowler should bat.

(a) THE CHOICE OF BOWLING END

The captain's decision as to which fast bowler bowls at which end is determined primarily by the wind. Because the out-swinger is the key attacking delivery, it is natural to give the better out-swing bowler the help of a wind blowing from leg to off. If the other bowler bowls in-swing, or is a left-arm bowler, so much the better.

However, the decision may not be so simple. If one bowler is much faster, the extra speed itself may be the key to an early wicket. If so, that bowler may be most effective with every bit of assistance from the wind, even if it is angled in a way not helping the most likely swing from either end. In addition swing often is more effective when the ball is bowled into the wind than when it is bowled with its assistance.

Therefore a captain with an especially good 'into the wind' bowler might choose to bowl that person into the wind regardless of pace.

Because dismissals win matches, the captain's decision must give the most likely wicket-taker the end that maximizes the chance of an early wicket. The person who should bowl first is the one who can exert pressure from sheer speed, swing, or precise control of length and direction from the outset. Therefore the new ball may best be given first to the opener with the best control, even if another is faster. At all levels the 'feeling' about a match and the conditions, in the minds of the openers and the batsmen waiting to follow them, can be affected to the advantage of the bowling team by an opening over of immaculate control and attack with a menacingly attacking field.

(b) WHERE THE FAST BOWLER SHOULD FIELD

Because fast bowling is very tiring, the bowler needs to recuperate between overs. The worst positions for rest and recuperation are the non-stationary positions in the arc from point through mid-off and mid-on around to square leg. In these positions a fieldsman has to sprint off the mark in one direction or another, dive for catches, bend down quickly to gather in the ball, and chase the ball towards the boundary when it has gone past.

Far better are the boundary-riding positions behind the wicket, or indeed anywhere on the boundary if a spinner is operating. Here the ball comes less frequently and in a way requiring less explosive movement. Also, the fast bowler's typically strong arm can be used to advantage, while these remote spots also facilitate 'switching off' and assist the avoidance of mental fatigue.

The slips might be thought ideal because one doesn't have to run at all and a few fast bowlers (Botham, Alderman) have excelled there. Such success is probably a rare thing in senior cricket, because a panting and perspiring fast bowler might not have the relaxed yet lightning reflexes that are essential, or because the intense concentration doesn't help recovery for the next over. Nevertheless, fast bowlers must work on their fielding skills in order to give the captain flexibility in being able to place the field.

(c) SETTING THE FIELD

There are some constants about field-settings for an effective fast bowler in a regular (not a limited over) match.

There will be at least two slips, a gully or thereabouts, a point or thereabouts, a third man or thereabouts, and a fine leg or thereabouts. That totals eight including the keeper and the bowler.

The remaining three are disposed according to circumstances such as whether the bowler bowls in-swingers or out-swingers, the state of the pitch, the shots favoured by the batsman, and the weaknesses being revealed by the batsman.

The knowledge of both the captain and bowler regarding these factors grows as the overs go by. The field is likely to be changed from time to time and from batsman to batsman. This is best done by the captain so that the line of authority and control on the field is not blurred. If the bowler wants a field change, the captain should be asked or communicated with by sign language if distance or the noise of the crowd prohibits talking. If the captain wants a major field change it is desirable that it is mentioned to the bowler beforehand. This is more likely to enlist the bowler's enthusiasm and support, even if the change is to plug a gap or for some defensive purpose required by the batsman gaining an advantage over the bowler.

By way of a warning to inexperienced captains—repeatedly shifting a fieldsman to the spot where the last boundary was hit is generally both unproductive and disheartening to a bowler who is bowling to a plan. Most successful plans have to allow for the occasional lucky or accidental shot and for the occasional mastery of the bat over the ball.

(d) WHERE THE FAST BOWLER SHOULD BAT

Some fast bowlers have been superb all-rounders, such as Botham, Kapil Dev, Davidson and Miller. Irrespective of how good a batsman a fast bowler may be, it is rare to find one batting at number one, two or three. As with wicketkeepers, fast bowlers are likely to be very tired at the end of the other team's innings and in need of a rest before being expected to play an innings of substance.

Therefore, while it is generally not good captaincy to rely on fast bowlers to bat at the top of the order, they have to work on their batting. Many matches are won by handy partnerships down the order and if selectors must choose between two fast bowlers of equal ability, batting and fielding ability will be taken into account.

CHAPTER 4

The Art of Swing Bowling

DARYL FOSTER and BRUCE ELLIOTT

Before discussing modifications to basic fast bowling technique it is imperative that the young fast bowler remembers the following two important points:

- **Strive to maintain 'good length' which is defined as that spot on the pitch that is just too short for the batsman to play forward comfortably and not short enough to play back. If the length is too short the chance of swinging the ball is also reduced. When striving for accuracy, aim to bowl in the 'corridor of uncertainty' which is an area from the off-stump to 20 cm outside this stump.**

- **The basis of effectiveness in fast bowling is speed. If a young fast bowler has the gift of being able to bowl fast then concentrate on developing speed. Line and length can be developed with experience and confidence.**

There is, however, more to the art of fast bowling than bowling fast with good line and length. The ability to deviate the ball from leg to off (out-swinger) or off to leg (in-swinger) as it approaches a right-handed batsman are skills that will enhance a fast bowler's repertoire.

Before discussing the grip or technique a bowler should use to make a ball deviate, it is necessary to explain *why* deviation occurs. It is all a matter of aerodynamics, because quite a small difference in the air pressures on the two sides of the ball will create an appreciable movement in flight. With a new ball, which is shiny on both sides, the elevation of the seam, angled to first slip for the out-swinger and to leg slip for the in-swinger (right-handed batsman), influences the flow of air on that side of the ball. This causes an air flow which is different on the two sides of the ball and produces a sideways force which results in the ball swinging. Note that the ball must be bowled so that the seam points in the same direction throughout the flight. When the ball loses its natural shine the bowler should polish only one side and bowl with the seam in a vertical position. The smooth layer of air on the shiny side now breaks away from the ball faster than the turbulent layer formed on the rough side, and the ball swings in the direction of the rough side.

It is also important to differentiate between early and late swing. The ability to swing the ball 'late' is one of the hallmarks of a good fast bowler. Generally, as a bowler increases the ball velocity, the air resistance drag increases proportional to

Figure 4.1: Air resistance with a new ball ... and with an old ball

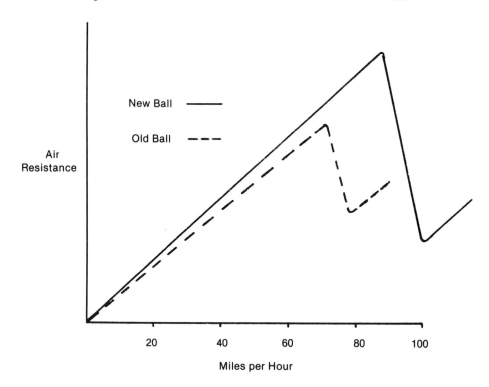

New Ball ——————

Old Ball — — —

Air
Resistance

20   40   60   80   100

Miles per Hour

the square of the velocity of the ball. When a cricket ball doubles its velocity from 18 m/s (64 km/h) to 36 m/s (128 km/h) the drag is not just doubled (×2) but is quadrupled (×4). However, at a critical velocity of approximately 40 m/s (144 km/h) the new ball suffers a sudden reduction in the drag to about one-quarter of the value previously recorded, thus allowing it to retain a high velocity.

What happens with late swing is that when a ball is bowled at or above the speed at which the critical value air resistance occurs, resistance will have only a small influence on velocity. However, when the velocity does eventually fall to below this critical level (usually relatively close to the batsman) there is a sudden increase in the effect of air resistance, and minor changes in pressure on one side of the ball compared with the other cause a small deviation in the trajectory of the ball.

A fast bowler may gain wickets through sheer express speed, even without the control to obtain late swing. However, if an aspiring fast bowler is unable to bowl at express speed, then the ability to swing the ball is an absolute necessity. Of special importance in the art of out-swing and in-swing bowling is the position of the hand and wrist at release. There is a need for a very precise hand-and-wrist action with the wrist remaining behind the ball, thereby enabling the maximum amount of backspin to be imparted to the ball. The position of the ball in the fingers also plays an important role in being able to swing the ball. The ball should be held in the tips of the fingers with the seam upright for late swing.

1. Out-swing Bowling

Genuine, late out-swingers are best delivered with the bowler usually using a side-on action. The aim of the right-handed fast bowler, bowling an out-swinger to a right-handed batsman, is to pitch the ball at middle-stump so that the batsman will have to play at each delivery as it swings away to outside the off-stump, just in case it might carry straight on. Bowling an out-swing delivery that deviates to outside off-stump increases the possibility of a snick to the keeper, to the slips or gully fieldsmen. To assist the delivery action, the seam should be just angled towards the slips with the arm action slightly beyond the vertical and finishing outside the left leg (right-handed bowler) in the follow-through. The bowler should also deliver the ball from close to the stumps.

Figure 4.2: Grip for out-swing delivery

2. In-swing Bowling

In-swing bowling requires a different approach than when bowling for speed or for an out-swing delivery. The ball is delivered from wider on the crease and aimed outside the off-stump so that it will deviate back to hit the stumps. The arm action is higher than for the out-swing delivery, with the bowling arm finishing down the right-hand side of the body in the follow-through. The complete action is therefore more open than that required to bowl the out-swing delivery.

As bowlers become more proficient they should aim to bowl the out-swinger and in-swinger with only slight modifications to their action. There are fast bowlers playing international cricket at the moment (e.g. Malcolm Marshall) who bowl the out-swinger delivery using an open body action. Control of the hand-and-wrist action enables mastery of this delivery with only minor modification to the normal action. Small changes in force applied to one of the upper quadrants of the ball by a wrist that is not 'square' to the batsman at release can be used to cause the ball to swing. However, the less the change in the position of the hand, the greater difficulty the batsman has in detecting the type of delivery to be bowled. Bowling to plan and varying the swing of each delivery are part of the psychology of bowling, but a young bowler must endeavour to master one of the fast bowler's stock deliveries, the out-swinger. This delivery, bowled with a side-on action, is still the best wicket-taking ball in cricket. Variations and subtleties are introduced gradually after mastery of the stock delivery and they should not be overused.

Figure 4.3: Grip for in-swing delivery

3. Cutting and Seaming the Ball

A great fast bowler swings the ball through the air and then also cuts the ball off the wicket. The 'cutter' is much better suited to being bowled on softer wickets such as are more frequently found in England. Because the cutter deviates in direction after hitting the wicket it is much more difficult to play than a ball that can be watched over a longer period. In Australian conditions, where hard wickets abound, cutting the ball is more difficult. The action involved in cutting the ball is difficult to learn and requires a great amount of skill and net practice. The leg-cutter grip requires the middle finger to be placed adjacent to the seam with the index finger spread as wide as possible (Figure 4.4B). The ball sits further back in the hand than usual, with the hand and fingers coming down the off-side of the ball on delivery so as to impart the leg-cut rotation.

The opposite grip applies for the off-cutter, with the hand and fingers cutting down the on-side of the ball (Figure 4.4A). Both off-cutters and leg-cutters should be bowled at the middle-stump and off-stump so that if the ball does not deviate to any extent, then LBW or caught behind the wicket decisions may still be achieved against the batsman. The cutting delivery is difficult to bowl on a hard wicket and with a new ball. If the wicket is not giving the bowler any assistance then one should forget about trying to cut the ball and concentrate on swing. Great patience and control are required to master the skill of bowling a cutter while retaining the ability to swing the ball. The rotation of the hand required to 'cut' the ball is completely different from the bending (flexion) action at the wrist required to swing the ball.

Therefore, always complete a net session by practising the in-swing and out-swing deliveries so that their basic movement patterns are imprinted on the mind.

Figure 4.4: Grip for off-cutter and leg-cutter deliveries (right-handed bowler and batsman)

A: Off-cutter   B: Leg-cutter

'Seaming' the ball is different from 'cutting' the ball. The ball must land on its seam in a 'seaming' delivery so that a minor deviation to the off-side or on-side occurs as a result of the impact with the wicket. Mostly, the bowler has little idea which way the ball will seam. For a seaming delivery, the ball is held upright (with the seam vertical) and a little further back in the hand than for the normal delivery (Figure 3.1). It is usually bowled using an out-swing action. When attempting to bowl seamers, it is necessary to project the ball with high speed such that it lands on the wicket with the seam upright.

4. Variation is the Key to Success

Variety is often the key to becoming a successful fast bowler. The ability to change pace, use the crease to change the alignment of the delivery, or bowl a yorker followed by a quick bouncer are all part of the fast bowler's armoury. These deliveries need to be subtle variations in the basic strategy used by the bowler. The ability to hit the top of the off-stump or bowl in the 'corridor of uncertainty' just outside this stump with 90% of deliveries is a most important skill. After three or four deliveries of a good line and length, when the batsman is lulled into a false sense of security, it is timely for a different type of delivery to be bowled. This variation must be controlled by the bowler and should not just be a reaction to a particular situation such as being hit to the boundary.

The slower ball can be delivered using a variety of methods but should not be overused. The different ways of gripping the ball for the slower delivery include the palm ball (Figure 4.5), the one-finger grip (Figure 4.6), and leg- and off-break spin-

Figure 4.5: Palm ball grip   **Figure 4.6:** One-finger grip

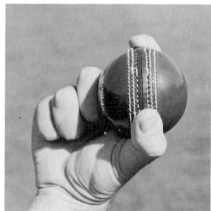

(the ball is placed further back in the
hand—see Figure 3.1)

bowling action. It is a matter for personal preference, and trial and error on the part
of the fast bowler as to what grip should be adopted. As all of these grips are sub-
stantially different from the normal method of holding a ball, the grip for the slower
ball should only be adopted at the last moment.

The yorker is another delivery that must be mastered. It is usually a fast, well-
pitched-up delivery bowled with the aim of passing under the bat and hitting the
base of the stumps. It is a good delivery to bowl before a batsman is set or to a batsman who uses a high back lift.

The bouncer is a legitimate delivery pitched short of a length aiming to pass the
batsman between chest height and somewhere above head height. The bouncer needs
to be bowled faster than usual because the ball loses speed on leaving the wicket. The
line the bouncer should take depends on the batsman's strengths and weaknesses,
but generally over the top of off-stump and middle-stump is an effective line. The
delivery needs to be practised, and bowling into an empty net to an appropriately
placed marker on the back net is one effective practice drill.

All these deliveries need to be thoroughly practised and mastered in the nets
before being used in a match situation. The young fast bowler should concentrate on
mastery of the basic deliveries and gradually add these other deliveries to the
armoury. It may take a long time to master just a couple of these additional
deliveries, but effective use of net practice can assist in this learning process. A part
of every net session should be spent practising different deliveries. Fast bowlers must
remember they are not just 'workhorses' for the batsman during a training session
and they must strive to improve their wicket-taking ability.

CHAPTER 5

Psychology of Fast Bowling

SANDY GORDON

The psychology of fast bowling refers to the mental skills required to perform this activity. Although rarely considered by the bowlers themselves to be as important as the physical, technical and tactical components of bowling, the acquisition of mental skills often can differentiate between First Class/Test-level bowlers and Club- or League-level bowlers. All other things being equal—talent, a good work ethic and adequate opportunities—bowlers who have learned or developed their mental skills, albeit often unconsciously, will be consistently more successful than bowlers who have not developed these skills.

Coaches and players who question the value of a mental skills training programme are able to produce records of legendary bowlers who neither were formally taught mental skills, nor appeared to need them. The sceptics' reliance, however, on 'experience' or the old adage 'If you've got it, you've got it . . . if you haven't got it, well . . .' ought to be challenged. For example, all cricketers can improve their ability to concentrate and cope with stressful situations. Furthermore, a mental skills programme short-circuits 'experience' by promoting strategies that will teach individuals, from a young age, how to deal with debuts or competing against top teams and difficult opponents. One could ask any cricketer how important 'state of mind' is in achieving success by simply having him or her decide what percentage of his or her game is 'mental'. If the response is compared with the actual percentage of practice time devoted to training mental skills then the disparity between the two figures is likely to be alarming. This should provide enough incentive for fast bowlers and coaches to take time to learn and practise a mental skills programme designed specifically for fast bowlers. The case for mental skills training in today's world of élite sport is quite convincing.

1. Demands of Fast Bowlers

(a) TYPE OF ACTIVITY

Fast bowling is often classified as a 'repetitive, independent closed-skill'. This jargon means that bowlers repeat essentially the same skill, with only subtle changes, over and over again almost as if they were all by themselves on the field. Bowlers are indeed uniquely in control of performance outcomes and their performance environ-

43

ment is relatively unchanging and stable. It is vital to understand these basic features of fast bowling.

(b) KNOWLEDGE OF INTERRELATED FACTORS THAT INFLUENCE CLOSED-SKILL PERFORMANCE

At least five interrelated factors can influence the performance of a fast bowler:

- biomechanical or technique factors;
- physiological conditioning or fitness factors;
- strategy or tactical factors;
- environmental factors such as climatic and wicket conditions;
- mental or psychological factors.

It is often too easy for bowlers to interpret weaknesses in performance as being due to psychological or mental factors. For example, a bowler may attribute poor bowling to lack of concentration (a mental factor) when the out-swinger is poorly performed (technique factor) or when the bowler is fatigued (fitness factor) or when this type of bowling is played well by the batsman (strategy factor).

2. Mental Skills for Fast Bowlers

While élite, professional fast bowlers are very aware of the mental demands of their trade, their traditional means of coping with these demands can be somewhat naïve. The skills presented in this section represent practical and useful mental skills reported by élite fast bowlers themselves. Space does not permit a full description of each of the six mental skills, but the nature and significance of these skills and brief practical suggestions for applying them are provided.

(a) GOAL SETTING

'If you don't know where you are going ... you will probably end up somewhere else!'

Why important?   Bowlers covet miserly figures, averages and strike rates of others and spend a lot of time wishing they could achieve the same. An effective goal-setting programme directly replaces wishful thinking with identifiable 'do' strategies that directly enhance performance. Having clear goals and clear strategies facilitates:

- motivation, confidence and determination;
- an understanding of the aspects of bowling that need to be performed as well as the development of new strategies to learn them;
- the will of the individual to make initial efforts to improve and to persist with training and practice.

How to practise.   Bowlers are mostly concerned with performance goals, such as 'how do I improve my bowling figures?'. Few, however, realize how important other areas of their lives and athletic preparation could be in their quest for excellence. Fast bowlers ought to set goals in at least five related areas.

EXERCISE: Copy these five goal areas into a diary and **write down**:

- your personal goals for each area;
- your strategies to achieve these goals in the form of answers to the same set of six questions for each area.

PERFORMANCE GOALS—e.g. to improve my strike rate (incorrect goals are: 'be better than last year', 'be the best').

1. What do I have to do?
2. When do I want this to happen?
3. Who can help me achieve it?
4. How do I go about it?
5. How can I measure it?
6. How will I know I have achieved it?

PHYSICAL GOALS—e.g. to improve my aerobic fitness (endurance) (incorrect goals are: 'get fitter', 'get faster, get stronger').

ENVIRONMENTAL GOALS—e.g. to structure my weekly activities so that I can include time every day with my family (incorrect goals are: 'manage my time better', 'get organized').

BEHAVIOURAL GOALS—e.g. to be more co-operative and positive (incorrect goal: 'develop a better attitude to the game and the team').

MENTAL SKILLS—e.g. to acquire and practise visualization skills (incorrect goal: 'be more mentally tough, and psyched up').

N.B.   Common mistakes to be avoided, as illustrated above, include setting goals that are too general and sometimes unrealistic. The following steps should be considered when setting goals.

Goal-setting Guidelines

SET difficult but realistic goals and state specific goals in measurable and behavioural terms. For example: 'I'm going to average one wicket per ten overs by mid-season and train two times a week for one hour'. Include short-range as well as long-range goals and performance-related goals versus outcome goals. The latter goals include such concepts as bowling line and length as opposed to just planning to take wickets. There is a need for goals to be set for practice and for games, and they should be positive rather than negative goals. It is preferable to be positive and aim to 'bowl line and length' rather than negative and think 'I mustn't bowl off-line'.

IDENTIFY target dates for attaining goals, and strategies for achieving the goals. Seek consultation with support personnel, including the coach, trainer, sport psychology consultant, senior players and other advisors.

RECORD goals once they have been identified.

PROVIDE for evaluating goals with periodic checks of progress by oneself and others. Organize support for achievement of goals from the family, coaches or team mates who can assist and encourage one's efforts.

Summary

Goal setting is the most effective 'mental skill' for facilitating both motivation and performance because it structures the bowler's efforts over time. Goals should not be committed to memory but written down and checked periodically. Each week one checks if actual behaviour is consistent with goals that have been set in all five achievement areas.

(b) CONCENTRATION AND ATTENTION SKILLS

'Sport intelligence is paying attention to the right things at the right time.'

Why important?   Fast bowlers have to attend to deciding on the next delivery and how to execute it, dealing with stressful situations that occur during play and thinking about field placements. These circumstances require an ability to shift attention from time to time and adopt an appropriate focus for attention at any one time. While some players are able to shift attention successfully, most experience at least an occasional blank period of concentration during which they are not totally aware of what they are doing. They simply run in and 'bowl fast' with no specific focus. At the other extreme, some bowlers may concentrate so intensely during play and in between overs that they mentally exhaust themselves prematurely. The key to effective concentration therefore, lies in selective attention and knowing how to 'switch channels' in concentration from 'off' to 'on' at appropriate times during play.

How to practise.   Selective attention is a most important characteristic of successful fast bowlers. It is important to learn how to reduce irrelevant and distracting cues to a minimum during performance. In essence this means:

- focusing on one thing at a time, such as targeting on the wicket just prior to delivery. Focusing on anything else at this time is poor selection and will probably distract the bowler;
- remaining in the here-and-now (present). Concerns about previous bad overs or the previous bad ball (which may have been hit for six), or future overs after lunch or tea, are counterproductive. Bowlers have no control over either past or future events and therefore all attention should be directed to matters over which they do have control, namely the very next moment in their lives which is the next delivery.

Switching 'on' and 'off' during performance requires learning how to 'switch channels' in attention and concentration. This is an important skill that requires practice. By 'switching on', bowlers optimize concentration around the point of delivery. By 'switching off' after a delivery and between overs or bowling spells, bowlers conserve energy and create opportunities to re-energize themselves.

A pre-delivery routine: switching on should include a 'checklist' of activities and a 'trigger'. The checklist might include habits or rituals like polishing the ball a certain way, or a pre-approach scratch or skip step; and 'do statements' consisting of mood or cue words, such as 'begin again', 'steady', 'smooth', 'control', 'explode'. These actions and self-statements serve to funnel the bowler's attention gradually from a broad to a narrow focus of attention and precede the trigger mechanism which finally shifts attention on the single most important cue prior to delivery, which is the target. Triggers might also consist of actions (e.g. last stride or look over shoulder) or they may be statements (e.g. 'target' or 'now'). Whatever means is used to achieve the trigger, the effect must be a signal to the bowler to zero in on one final cue only.

A post-delivery routine to help bowlers switch off should also make use of physical and mental devices. Deep breathing, shoulder shrugs, stretches and shadow bowling (bowling without the ball) are all useful physical devices for relieving tension. Focusing briefly but intently on something in the field (e.g. the colour of the grass) or at the cricket ground (e.g. a billboard) or someone in the crowd are all useful mental devices.

Summary

Concentration skills demand patient and persistent practice. Learning to switch channels effectively during performance requires selective attention, and the development and active practice of pre-delivery and post-delivery routines. Even small improvements in concentration over time can have dramatic effects on bowling performances at all age and competitive levels.

(c) ANXIETY AND PRESSURE: STRESS MANAGEMENT

'What happens to you is nowhere near as important as how you react to what happens to you.'

Why important? Mental toughness is a characteristic ascribed to bowlers who perform well under pressure and in adverse circumstances. In these situations such players are able to keep their mental and physical arousal levels within manageable limits, which in turn facilitates performance. Adversity in fast bowling is a highly subjective experience but it might occur when the wicket is flat and unresponsive, or when one is tired but forced to persevere, or becoming increasingly frustrated by batsmen playing and missing, team mates dropping catches and/or questionable umpiring decisions. Playing away from home or in front of a hostile crowd, and playing when injured are also examples of adverse circumstances. Unfortunately, under these conditions bowlers may become inwardly angry and give up. Sometimes they become visibly angry, thereby losing emotional control. On other occasions they become hostile or panic stricken and end up trying too hard, which acts only to compound the adversity.

All the above responses are unnecessary and detrimental to performance. These adverse circumstances are not inherently stressful *unless* bowlers choose to perceive them as such. Bowlers, therefore, can learn that pressure is really something they put on themselves. It can therefore be controlled by integrating both physical and mental coping skills.

How to practise. Physical techniques that help bowlers control and manage pressure and problem situations immediately are:

- At least three deep breaths. This may sound too simplistic but if bowlers stand, or walk slowly, and focus on the sensations of deep breathing, their focus and attention immediately internalize to body processes. This provides a momentary welcome relief from external stresses.

- One or two deep release breaths. Again, this simple technique of inhaling deeply, then forcibly exhaling, effectively relieves tension. This is particularly so when bowlers feel rushed or hurried.

- Muscular relaxation. By alternately contracting and relaxing certain muscles and muscle groups which feel tight and stiff from prolonged periods of inactivity or overuse (e.g. upper- and lower-limb muscles, shoulder and back muscles), relief from physical stress is achieved. Coupled with passive stretching exercises, muscle relaxation techniques are excellent for dealing with physical tension.

Mental techniques such as 'thought stoppage' are also very effective and easy to use. Thought stoppage has three simple steps:

- Bowlers first learn to listen to their inner voice, which is what they say to themselves and what they are thinking; and to become particularly sensitive to any negativism such as 'if only I hadn't delivered that last ball . . .'

- Bowlers should immediately halt negativism and stress-producing dialogue, by saying to themselves 'STOP!' Cricketers who use this technique are amazed at how effectively this actually stops counterproductive and dysfunctional self-talk.

- Bowlers should then replace negative self-talk with statements that are functional, positive and task-oriented such as '. . . next time focus on the top of the off-stump, begin again'.

The most simple, helpful and probably most important 'mental device' for dealing with anxiety and stress is a focus on task-oriented behaviour.

When confronted with a problem, bowlers should immediately ask themselves one question: '. . . What is it I have to do, right now?'. By sticking with this strategy and with the process of bowling (i.e. how to bowl), and not the product (i.e. results and winning), bowlers will work their way through difficult situations without panicking or losing control.

There are other effective mental skills to help bowlers deal with stress. As with thought stoppage, the basic premise is always to prevent negativism from inducing or adding to the pressure. Pressure is averted by employing an inner dialogue which is task-oriented and characterized by positively worded 'do statements'. A useful exercise for all bowlers, illustrated as follows, is first to prepare a list of stressful situations. Opposite this list prepare an integrated mental and physical coping response for each situation. For example:

WHAT IF ... Stress situation	DO THIS ... Integrated coping response
1. I'm being hit all over the ground in a limited-overs game.	1. SAY 'next ball ... in the block hole' 'keep your rhythm and line' 'this is the ball ... c'mon' 2. DO 'deep breaths, shoulder shrugs' 'visualize the next ball' 'pre-delivery checklist and trigger'

Summary

An approach that combines both mental and physical coping skills can assist bowlers in keeping arousal levels within manageable limits. Bowlers can prepare pro-actively for stressful situations and should learn that these situations are more 'creations' than 'fact', and that difficult circumstances ought not to be perceived as threats but as challenges to be overcome.

(d) VISUALIZATION SKILLS

'Visualization is one of the most powerful mental training strategies available to all performing athletes.'

Why important? Research results and the reports of élite performers in closed-skill sports such as golf and gymnastics clearly show the incredible power of mental rehearsal and mental imagery in both learning and enhancing sport skills. It would be safe to say that while there is no substitute for active bowling practice, bowlers who both practise and perform vizualisation skills will achieve higher levels of performance than those who merely practise. Vizualisation works, and bowlers who do not use this particular mental skill in both practice and games are not maximizing their potential.

How to practise. The following critical features of practice should be followed for successful visualization:

- Visualization skills are best learned while in a relaxed state. Bowlers should therefore first practise at the nets by relaxing with, say, deep breathing and then vizualise the type of ball they wish to deliver. Later, visualizsation can be built into pre-game and during-game strategies.

- Ultimately, visualization should be of an actual performance at an actual venue. Features of the environment should be noted so that the visualized performance is realistic and vivid.

- In the visualization, the bowling action should be performed to include all phases from the run-up to follow-through and, if preferred, against a particular batsman. Performing the complete action in its entirety ensures proper sequencing of actions and total perspective for the performance.

- Only successful bowling images should be used. Unsuccessful or poor images increase the likelihood of errors. Thinking and seeing a positive image is infinitely more powerful than merely thinking positively. Positive thoughts follow naturally from positive images and not vice versa.

- Repeat the visualized bowling as often as necessary. Three or four repeats may be all that time will allow in between deliveries during an over. However, ten or twelve repeats while waiting in the field is advised.

- Visualize at the real speed and not in slow motion. Timing and rhythm may be adversely affected as a result of slow-motion images which might create inappropriate effects. All bowling actions should therefore be visualized at the correct speed and under competitive circumstances.

- As well as 'seeing' the technique, the 'feel' of the action ought to be vividly sensed and experienced. Harnessing these kinaesthetic cues in the image will make the visualization more powerful. For example, the feel of the ball in the hand, the feet floating across the turf, the speed of approach, the arc of the bowling arm, the perfection of the release of the ball and the follow-through can all be imaged.

- Finally, a key element in successful visualization is intentionality. Bowlers must have a true desire to have or create the bowling action they wish to visualize. They must also believe that it is really possible to attain such results.

Summary

Choose a positive and successful image of a desired bowling action and replay this image ten to twelve times vividly, 'in colour', at the wicket or ground, and enjoy the sensations of confidence and success that are created. If the bowler's intention is firm, the total technique should begin with relaxation and then visualization in the order of the steps described above.

(e) CONFIDENCE AND CONSISTENCY

> *'If you've made up your mind that you can't do something ... you're absolutely right!'*

Why important? Put simply, confidence affects bowling performance, and performance affects self-confidence. Consistency of results therefore directly affects a bowler's self-confidence. Most cricketers find confidence and consistency elusive. They are generally puzzled as to how they can be consistent, and stay on the winning and self-confidence spiral which breeds success; while staying off the losing and increased diffidence (lacking in self-confidence) spiral which spawns poor and erratic performances. The key to both confidence and consistency is effective skill in setting goals.

How to practise. Self-confidence is concerned with neither what fast bowlers hope to do, nor what they have already achieved. It is concerned with what bowlers realistically believe they are capable of doing at a particular moment. In other words, self-confidence involves a belief in current ability. While diffident bowlers quickly lose a belief in their ability, over-confident bowlers may delude themselves with false confidence. The confidence may be false because they believe they are so gifted that they do not prepare themselves or practise as effectively. In both cases performance deteriorates.

To enhance confidence and consistency bowlers should do the following:

- Prepare and develop a consistent thought pattern and focus. Pre-game preparation (physical, technical and mental) will facilitate consistent results. Visualization, combined with self-talk, ought to be integrated with active physical preparation.

- Be positive, determined and committed. Bowlers should choose to feel and be positive in what they do. This helps to ensure that they perform within their capacity and are in control, both mentally and physically.

- Mean what is said. Self-talk about expectations for performance will have a corresponding effect on the body only if stated with emotion and conviction. How one walks, talks, listens and looks reflects the person. If one really means it when saying 'I feel 3 m tall and bullet proof' or 'I own this place', the changes in walking, posture and especially the outlook on events can be observed.

Bowlers lacking in confidence often appear 'frightened' or they 'play scared'. They seem to lack fluency and smoothness in their action, and rarely appear to be having fun. Diffident bowlers should consider the following:

- Not be prisoners of negative images. Bowlers who lack confidence tend to focus so intently on mistakes and errors that it distracts them from attending to those aspects essential for good performance. As in both the goal setting and coping skills sections previously presented, bowlers should ask themselves, 'What is it I have to do?' rather than, 'What is it I'm doing wrong?'.

- That's not like me'. When out-of-form bowlers are having a lean period they typically discount the positives in their performances and conjure up images of themselves being far less talented and less skilled than reality. Good bowlers never lose their ability and skill but they can lose confidence in their ability and skill. Hence, out-of-form bowlers should deal only in successful images and best performances, and stop dwelling on uncharacteristic, less talented images.

- Think realistically about performances. Bowling performance is rooted in realistic expectations and beliefs. Some bowlers actively look for 'proof' that they are out of form and evidence to support their belief that they 'just can't do it'. This must be short-circuited before a negative, self-fulfilling prophecy develops. This is especially true in the first few matches after a promotion, or even the first few overs, if they have not been successful. Bowlers should therefore expect to do well and look instead for proof that they 'can do it'. Paper cuttings and videos of previous successes are excellent examples of proof. This intervention does not deceive bowlers, it simply reminds them that their ability has not disappeared.

Summary

Effective goal setting will arrest the downward spiralling of self-confidence that bowlers frequently create as a result of temporary lapses in form. By remaining task-oriented' in both thought and strategy, and realistic in expectations of themselves, bowlers can work through temporary lean spells. Sometimes there is no particular reason why bowlers perform poorly; some sport crises simply cannot be explained. What is critical, however, is that bowlers react immediately to loss of form and do everything they can for themselves to restore and sustain confidence.

(f) GAME STRATEGIES

> *'A necessarily different mental or psychological state exists when a bowler is performing well as opposed to when he or she is performing poorly—it is called the ideal performance state.'*

Why important? Optimal preparation for games helps to produce optimal performances and this is made possible by analysing previous satisfactory or outstanding performances. Bowlers tend to analyse only poor performances because mistakes and errors draw attention to themselves. Good performances on the other hand, which provide excellent learning opportunities, may be rarely scrutinized with as much detail and concern. All bowlers, therefore, can help re-create their 'ideal performance state' by becoming more aware of how they feel and how they behave during good performances.

How to practise. Observations on the following are worth recording:

DAY BEFORE THE GAME:

How were daily activities structured? Rest and nutrition are particularly important. What did you do that you think helped?

GAME DAY: PRE-COMPETITION PLANNING

Prior to arriving at the field:
- wake-up procedures, structure of activities, nutrition and rest?
- control of arousal, maintenance of confidence, visualization?
- gear preparation, travel to field (did you drive or were you driven)?

At the field:
- field/wicket inspection, group and individual warm-up?
- practice of physical and mental activities?

DURING THE GAME

Where was your focus? Focus ought to be on individual and team mini-goals within each session and bowling spell, and a philosophy of 'ball by ball' delivery strategy.

POST-GAME ANALYSIS

The four Rs should be addressed:
- review what happened;
- retain what can be learned from good and bad individual and team performances, and what can be learned from the opposition (discard the rest);
- rest regardless of how well you bowled. It's over, so rest and re-energize;
- return to practice refreshed and eager, feeling positive and determined about future games regardless of previous performances.

Summary

An analysis or de-briefing of pre-game, during-game and post-game activities is required in order to help bowlers prepare effectively for future games. By recording patterns of feelings and behaviour during good bowling spells, an 'ideal performance state' comprising physical, technical and especially mental activities is fashioned. Bowlers should strive to re-create this optimal performance condition for each subsequent game.

3. Communication with the Fast Bowler

The fast bowler, during a string of overs, is by far the hardest worker on the field; experiencing more physical discomfort, more fatigue, and possibly more stress and pain than anyone else. The fast bowler is both the endurance and explosive athlete of the cricket team. Not all fast bowlers, under these circumstances, can maintain the proper positive attitude, and they may become discouraged or angry or in some other way fall into a mental attitude not conducive to the best physical endeavours.

One of the captain's jobs is to observe or sense any such problems and any negativity, and assist the bowler in recovery.

This can be done by the usual means of encouragement and praise. The 'football coach' model of yelling abuse might be productive in collision team sports but is almost never the best thing to do in cricket, given the special nature of the game and the special nature of fast bowling. If criticism from the captain is unavoidable, as it may be if there is no coach, it may best be made after a spell of bowling, after an entire session of a match, or indeed after the match.

However, if something readily observable has gone wrong without the bowler realizing it, like loss of rhythm, bowling from too wide of the stumps, or clearly gripping the ball wrongly, the captain may be able to quietly indicate whatever might be worthy of attention. Wicketkeepers can often pick those things up and all fast bowlers should strive to develop a working liaison with the wicketkeeper.

Notwithstanding these occasions when the captain must talk to the bowler, the captain should not do it too much or too often. The fast bowler has to concentrate on the next delivery and will have switched off other considerations and switched on to the prime concern well before commencing the run-up. Unnecessary chatter or repetitive remarks like 'Keep it up' can interfere with the operation of the bowler's own mental devices.

4. Conclusion

Collectively this outline of psychological techniques for fast bowlers represents a series of mental skills that will facilitate performance enhancement at all age and competitive levels. This is especially so if they are taught well by sport psychology consultants and coaches, and integrated with technical and physical skills. Fast bowlers who expect immediate results might be disappointed in the same way they may have been disappointed with their first efforts at bowling. However, with proper direction and persistence on their part, bowlers will eventually notice improvements in performance.

CHAPTER 6

Factors that may Predispose a Fast Bowler to Injury

BRUCE ELLIOTT, DAVID JOHN and DARYL FOSTER

A fast bowler can suffer a serious injury through an error in technique, from inadequacies of physique or physical preparation, from the extreme physical stresses of fast bowling or from a combination of all three.

1. Physical Attributes

The fast bowler should possess a high level of muscle power and muscular endurance to bowl fast. The muscle power is needed to move the limb segments at the desired speeds and muscular endurance allows the bowler to sustain top pace during the entire bowling spell. Sufficient flexibility is required to enable a complex series of body movements to occur smoothly through a full range of motion without excessive stress and strain on joints and associated tissues. Cardiovascular endurance (aerobic fitness) and the ability to repeat near-maximum efforts with each delivery (anaerobic fitness) are also needed so that ball speed and direction can be maintained for prolonged periods. The development and maintenance of these physical attributes are important if fast bowlers are to bowl to their full potential and remain injury-free.

(a) HAMSTRING AND QUADRICEPS MUSCLE-GROUP STRENGTH

It may sound strange but if the hamstring muscle group (back of thigh) is less than approximately half as strong as the quadriceps group (front of thigh), then the imbalance may predispose the bowler to hamstring and knee injuries. Bowlers often develop the quadriceps group to such an extent that the hamstrings, particularly if the flexibility of this muscle group is poor, are in danger of being strained or torn. This commonly occurs when the leg is thrust forward at front-foot impact. The stability of the knee joint is also threatened if the quadriceps muscle group becomes disproportionately stronger than the hamstring group.

(b) UPPER BODY STRENGTH

While shoulder region and upper-limb strength are needed to generate force, fast bowlers and coaches should realize that ball speed is derived from a precisely *coordinated* sequence of body segment movements. The fast bowler who relies on upper- body strength rather than a combination of strength and co-ordinated movement to develop ball speed may apply excessive twisting forces to the spine during the delivery action which could contribute to the development of back injuries.

(c) ABDOMINAL AND BACK STRENGTH

Weak abdominal muscles and lower-back extensor muscle strength will reduce the support for the spine during the complex series of movements involved in fast bowling. The fast bowler must develop the muscles surrounding these areas.

(d) FOOT ARCHES

The foot is designed to absorb impact forces during movement, but if the bowler has a low foot arch, then these forces may not be absorbed as effectively as would occur in a normal foot where ligaments provide greater support and resilience. With a low arch the lower limbs and lower back would be subjected to higher levels of force than may be the case for a bowler with a high or medium level arch. Bowlers with a low foot arch should therefore pay particular attention to the quality of their footwear and may need special inserts in their boots.

(e) POSTURE

Deviations from what is considered a normal posture are classified as postural deformities and often can be a primary source of injury. These can be due to skeletal misalignment or asymmetries caused by the development of one side of the body at the expense of the other. For example, a more muscular development of the bowling side of the body can occur when compared with the non-bowling side.

Figure 6.1: Asymmetrical body development

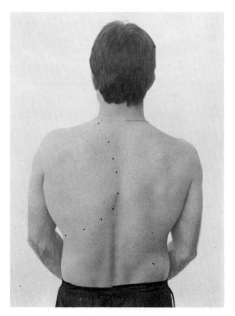

Aspiring fast bowlers with abnormal curvatures of the spine (sideways or front-to-back) must pay particular attention to the physical preparation aspect of their pro-grammes. The one-sided nature of fast bowling can cause atypical lateral curvature of the spine called scoliosis (Figure 6.2A). If scoliosis is already present, it will almost assuredly be aggravated by the repetitious fast bowling action. Abnormal hollowing of the lumbar region spinal curvature is called lumbar lordosis (Figure 6.2B) and it too can be a predisposing cause of lower back pain and injury.

The nature of bowling places excessive strain on the lower back, and can pre-dispose the fast bowler to lumbar lordosis. If either of the above back postures are evident then bowlers should seek professional advice to correct these functional dis-abilities.

Fast bowlers need to prepare their bodies well for the task at hand, namely bowl-ing fast, with minimal chance of injury. Remember that a fast bowler must train to play, not play to train.

Figure 6.2: Postural deviations often evident in fast bowlers

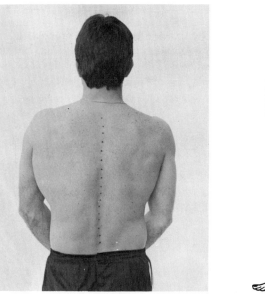

LORDOSIS

A: Scoliosis   B: Lumbar Lordosis

2. Technique

Research by Foster and associates has shown that certain body positions during the delivery stride can contribute to injury in fast bowlers.

(a) FRONT-ON ACTION

The current trend of bowling with a front-on action in preference to the classic side-on technique causes the lower back to hyperextend more during the delivery stride, particularly if the bowler tries to view the batsman outside the non-bowling upper limb. The combinations of hyperextension, lateral flexion and rotation of the trunk (Figures 3.3C and 3.3D) have been postulated as reasons for the increase in back injuries.This is particularly so with young players who have a foot and hip alignment common to a front-on action and yet still attempt to adopt a side-on shoulder alignment (see Chapter 3).

(b) SHOULDER ROTATION

The excessive rotation of the shoulders (the bowling-shoulder moves away from the batsman) in an endeavour to improve the side-on position of the shoulder alignment between back-foot and front-foot impact (Figures 3.3C and 3.3D) is a characteristic of front-on bowlers and has also been linked to back injuries. This rotation of the shoulders is then immediately followed by another rotation of the shoulders (as the bowling-shoulder is moved forward) toward the batsman in an attempt to develop maximum delivery speed. These rotations of the spine and/or lower torso place stress on the lower back, particularly the lower lumbar vertebrae. The West Indian fast bowlers generally use a front-on bowling action. They do not, however, significantly change their shoulder alignment from back-foot to front-foot impact. Neither do they hyperextend their back during the delivery stride.

It would therefore seem important for coaches to emphasize **a side-on bowling technique where the back-foot, hips and shoulders are all in the correct alignment rather than to encourage young players to bowl as fast as possible. If a front-on technique is desired, then these bowlers MUST NOT combine a front-on foot position with a side-on shoulder orientation.**

(c) FLEXION OF THE FRONT KNEE

Bowlers flex (bend) the front-knee joint at front-foot impact in an endeavour to reduce the effects of the impact force. Studies have shown that the knee joint flexes by approximately 10° from front-foot impact to ball release.

3. Physical Demands

A fast bowler must absorb the large reactive forces from the ground during each delivery irrespective of whether using the front-on or side-on bowling technique. At each foot-strike during the run-up, forces of approximately three times the weight of the bowler are generated. This increases until at back-foot impact and front-foot impact during the delivery stride a force of four to five body weights must be absorbed (Figure 6.3). Properly fitted boots are capable of cushioning some of these forces and a relatively soft approach area will reduce the effect that reaction forces have on the body. The total number of deliveries is also a major factor if the bowler is to remain injury-free, because of the cumulative effect of a long day of bowling fast, in addition to chasing leather while fielding.

Number of deliveries. Players who bowl for two or more teams (School/Club or Club/State or League) may be required to bowl at four practice sessions and two games per week. Nineteen of a group of thirty two bowlers (59%) who bowled in excess of the mean number of matches for the total group tested (82 bowlers) were

Figure 6.3: Typical ground reaction forces at front-foot impact

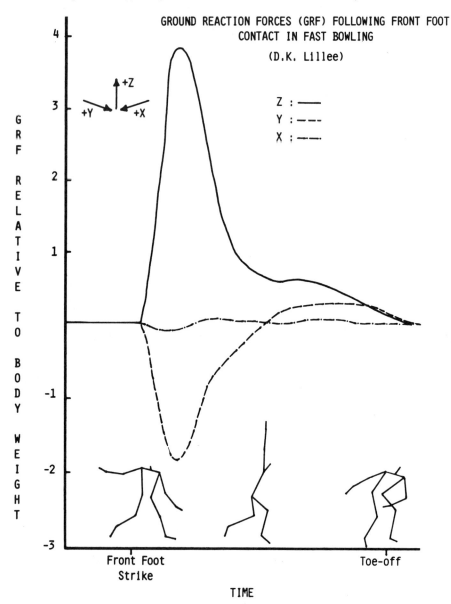

injured (nine stress fractures to the back; ten soft tissue back injuries) compared to the 38% injury frequency for the total group.

Foster and associates (1984) recommended a limit of three spells of six overs for bowlers under 19 years of age.

It would therefore seem evident that **bowling in too many practice games, or**

bowling too many overs in a single spell or repeated spells of the same game, may be a contributing cause of injury. It is important to reiterate the need to adhere to the following fast-bowling guidelines (Table 6.1).

TABLE 6.1

FAST BOWLING GUIDELINES

UNDER 12	A limit of 2 spells of 4 overs with approximately a one-hour break. Modified rules players may bowl less.	**2 x 30-min. practice sessions per week.** 5-min. short run—reduced pace 20-min. match speed—coach controlled 5-min. specific technique development
UNDER 16	A limit of 2 spells of 6 overs with approximately a one-hour break.	**2 x 40-min. practice sessions per week.** 5-min. short run—reduced pace 25-min. match speed—coach controlled 10-min. specific technique development
UNDER 19	A limit of 3 spells of 6 overs with approximately a one-hour break.	**3 x 40-min. practice sessions per week.** 5-min. short run—reduced pace 25-min. match speed—coach controlled 10-min. specific technique development
SENIORS	A limit of 3 spells of 8 overs with approximately a one-hour break.	**3 x 1-hour practice sessions per week.** 10-min. short run—reduced pace 40-min. match speed—coach controlled 10-min. specific technique development

(Reprinted from *Sports Coach*, Vol. 7(4), with permission of author and editor.)

CHAPTER 7

Common Injuries to the Fast Bowler

KEN FITCH

Because it can be described as a non-contact sport, the rates of injury to fast bowlers reported in recent years make this one of the most injury-liable non-contact activities. However, much can be done to reduce the incidence of such injury by understanding the causes and instituting appropriate measures both during training and playing (see Chapters 2, 3 and 4).

Most injuries sustained by fast bowlers occur from either overuse, poor technique or a combination of both factors. Environmental and developmental factors are other significant causes of injury.

Overuse injuries sustained by fast bowlers occur from too rapid an increase in:

- the amount of running undertaken as part of general training;
- the number of balls bowled both at practice and in matches.

Faulty bowling habits which may induce injuries include:

- failure to warm up and stretch;
- unsuitable or worn footwear;
- an excessively long run-up;
- a faulty (or injury-prone) bowling action;
- bowling fast, short balls ('bouncers') either too frequently or with a bowling technique which is unsuitable for delivery.

Examples of combinations of these aspects would be:

- an incorrect (or uncharacteristic) action caused by tiredness after bowling too many consecutive overs or too many overs in a day;
- an excessive number of balls bowled at or near maximum speed.

An environmental factor which may cause injury is the condition of the playing surface, especially around the bowling crease. Surfaces which are uneven, excessively worn or slippery may result in injury, the risks of which would be increased by unsatisfactory footwear. Cricket is a summer sport, and high temperatures, especially

60

if associated with high humidity, can cause dehydration and heat injury. An adequate fluid intake is essential in the prevention of such problems.

Developmental causes of injury include congenital defects of the lower spine. Body types which are unsuited to fast bowling are more likely to be associated with injuries. This is particularly so in adolescents who, despite advice to the contrary, continue to pursue their ambitions to be fast bowlers. Growth plates in the adolescent spine may fail from excessive stress with resultant damage and incapacity. It is vital that adolescent fast bowlers are not permitted to bowl more than the accepted guidelines (see Chapter 6).

There is a final group of injuries that will not be discussed in this chapter. These involve injuries which are sustained by fast bowlers while batting, fielding, throwing or playing sports other than cricket.

1. Injuries to the Back

The back is by far the most vulnerable region of a fast bowler's body. Injuries occur to the bones, discs, joints and muscles of the spine. Bone injuries are amongst the most serious and can occur from either overuse or a congenital (hereditary) or developmental condition. Since the stress fractures of the lower spine sustained by Dennis Lillee in 1973, much has been written about the condition. Like Dennis, other cricketers have been fully rehabilitated from their injuries and resumed bowling. Unfortunately, many have not been able to recover and have been either forced out of cricket, played only as a batsman or switched to slow bowling, such as Bruce Yardley. Because of the unique nature of stress fractures of the lower spine which may occur as a result of fast bowling, this condition will be discussed in detail.

(a) STRESS FRACTURES

The onset is typically associated with excessive bowling or poor bowling technique or a combination of both. Pain in the lower back on the non-bowling side, i.e. left lower-lumbar region in a right-handed bowler, is the first sign of a developing stress fracture. Initially, the pain is felt after bowling and soon disappears. However, characteristically, it will reappear when the player next attempts to bowl at or near maximum speed. At this stage, easy-paced bowling at the nets is likely to be pain-free.

Continued fast bowling will result in more prolonged and intense pain after bowling. This may continue in bed the night following a game. The next stage finds that the bowler is forced to stop bowling because the pain is too great to continue.

If medical advice is sought at this stage, the most distinctive feature is that pain is induced when the spine is extended and rotated. Such pain is located adjacent to the fifth (or possibly fourth) lumbar vertebra and this region is tender to pressure. Other movements of the lumbar spine may be pain-free or may cause only minor pain.

In the early stages of a stress fracture, these symptoms and signs customarily settle rapidly within a few days only to recur promptly with the resumption of fast bowling. A bone scan is likely to be positive (revealing the fracture) but an X-ray may be negative (fracture not visible). However, an X-ray may disclose some variation from normal in the spine, especially at the lumbosacral junction. A variety of structural abnormalities may be identified, the principal significance being that their presence

Figure 7.1: A lumbar vertebra stress fracture

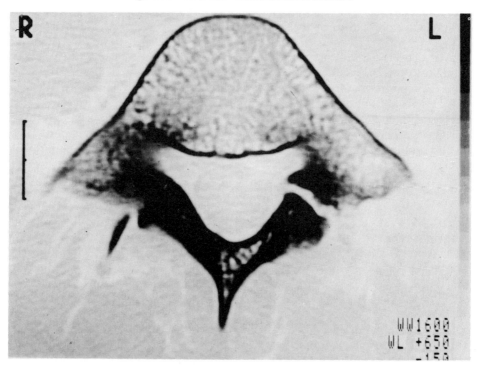

will tend to decrease the likelihood that the fracture will heal. A computerized axial tomograph (CT scan) is the most sensitive method of identifying stress fractures and following their progress (Figure 7.1).

The management of this condition involves the immediate cessation of bowling and the commencement of a rehabilitation programme (see Chapter 8).

If pain-free when running and batting, resumption of cricket may be permitted with care. However, bowling must not recommence until CT evidence of healing of the fracture has been obtained. The outlook for such a fracture appears to be related to the factors listed in Table 7.1.

TABLE 7.1

PROGNOSIS

Recovery is more likely if:	Recovery is less likely if:
L4 fracture	L5 fracture
History of overuse	No history of overuse
Sole structural abnormality	Associated spinal defect
	Family history of structural spinal defect
Conscientious rehabilitation	Poor rehabilitation

(b) STRESS REACTIONS

Symptoms which are typical of stress fracture but resolve promptly and are un-associated with any X-ray or CT evidence of fracture may be due to a *stress reaction*. In such circumstances, and provided the continued stress has been ceased, the bones of the spine react to the stresses of fast bowling in a positive manner and become stronger and better able to withstand such stresses. The X-ray and/or CT may show the pars interarticularis of the area prone to stress fracture as denser and whiter (sclerotic). In such circumstances, it is important to advise such bowlers to refrain from any injury-prone practices that may have contributed to the symptoms.

If an X-ray should reveal a major developmental defect of the lower lumbar spine, such as a spondylolysis (breaks in one or both sides of the neural arch), or spondylolisthesis (breaks on both sides associated with a forward slip of the vertebra just above the breaks) (Figure 7.2), the prospects of that player resuming fast bowling are poor. Nevertheless, this type of condition is trouble-free in a variety of other sports such as Australian football which do not require extension of the spine.

Figure 7.2: Spondylolisthesis

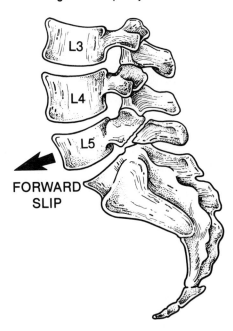

(c) OTHER SPINAL CONDITIONS

Discs. Intervertebral discs consist of semi-solid material (nuclear pulposus) surrounded by a semi-rigid covering (annulus fibrosus). A disc may bulge (and result in slight irritation of the coverings of the spinal cord) or prolapse (when the protruding disc material will compress the spinal cord and/or a nerve leaving the spinal canal). This results in pain and restriction of movement of the spine. Typically, forward flexion of the spine is reduced, with side bending also restricted. This is particularly

Figure 7.3: Prolapsed disc (L5)

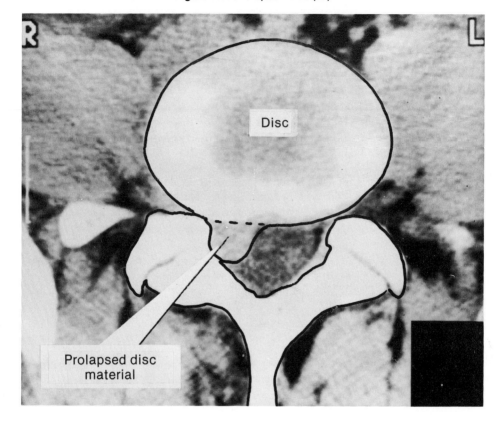

so if there is associated 'protective' muscle spasm. If a spinal nerve is irritated or compressed, pain will be felt in one leg or the hip and buttock region. When lying on the back, the ability to lift the leg on the affected side may be mildly or even severely restricted.

X-rays are generally unhelpful in diagnosing a prolapse but a CT scan is frequently successful (Figure 7.3). While a prolapsed lumbar disc may develop 'acutely' (suddenly), it is more common for symptoms to appear overnight or over one to two days. Bed rest is often essential in the initial stages, followed by a programme of swimming and back rehabilitation. Failure to respond to such measures, including physiotherapy, may justify more extreme and intensive measures such as an injection into or around the disc or even its surgical removal. Once fully recovered and rehabilitated, a return to fast bowling may be possible but it is essential that exercises to maintain good muscle support for the lumbar spine are continued indefinitely.

Joints. Injuries to the facet or 'joining' parts of the lumbar spine do occur in fast bowlers. A bowling action that causes overextension of the spine and downhill running undertaken pre-season can contribute to facet joint strains. The features of this condition are not dissimilar to those of stress fracture and stress reaction, with pain on extension and local tenderness. However, X-rays and CT scans would be normal, revealing nothing, and the condition responds well to rest and physiotherapy.

Muscles. Injuries to the muscles of the back are surprisingly uncommon in fast bowlers. Such injuries tend to occur as a result of failure to warm up and stretch, or from a strain caused by a faulty action when overtired. The onset is acute with pain and local tenderness at the site of the muscle which is torn. If the muscle is stretched or made to contract against resistance, pain is experienced. This condition responds well to physiotherapy in about three weeks.

2. Lower-limb Injuries

(a) FOOT INJURIES

Blisters from friction and callosities due to chronic pressure arise especially pre-season and early in the season. Poorly fitting footwear is often the cause, especially failure to gradually 'break in' new boots. Plantar fasciitis (inflamed tissue under the foot) is a painful and often disabling condition affecting the tissue (plantar fascia) beneath the skin of the sole of the foot. Although most commonly occurring beneath the heel, it can also arise in the arch of the foot. Inappropriate footwear, excessive running on hard surfaces and an abnormal arch (either too high or too flat) may be the cause and this may be aggravated by tight calf muscles. Rest, physiotherapy and calf-stretching exercises may bring about recovery but in many instances an arch support will be necessary. The assistance of a podiatrist is wise.

Stress fractures can occur in several bones of the foot but the second, third and fourth metatarsals are the most frequently injured. Too much running is the usual cause and is often compounded by excessive impact forces as a consequence of running on hard surfaces in shoes with insufficient shock absorbing properties. It can also arise from excessive bowling. Initially, pain in the forefoot occurs after running or bowling. It increases to restrict and then to prevent the player continuing to run or to bowl. Localized tenderness over the site of the stressed bone is evident and a bone scan will show the fracture. The condition requires six to eight weeks' rest from running and bowling followed by a gradual resumption. X-rays are negative for five to six weeks and the healing fracture is not always visible even during the late phases of healing. Maintenance of general fitness by swimming, cycling or running in water using a flotation vest should be undertaken during the recovery period (Chapter 8).

(b) ANKLE INJURIES

The sprained ankle is the single most common injury sustained in sport. Fast bowlers often succumb to this injury in the proximity of the bowling crease. Uneven surfaces are a major factor. The foot is suddenly inverted (twisted inwards), resulting in acute pain and swelling due to tearing of fibres of the lateral (outer) ligament of the ankle joint. There is immediate difficulty in bearing, or inability to bear, the weight of the body. The application of ice, compression, elevation and rest are essential and this is followed by physiotherapy. A precautionary X-ray is often wise in the more severely injured ankle, to rule out the possibility of a fracture. The period of absence from cricket will be longer for a more severe injury, but can be lessened by more intensive treatment. In addition to continuing the rehabilitation exercises, protective strapping is usually recommended in the early weeks or months after resuming cricket. Occasionally, an ankle ligament injury appears to recover but the victim is unable to run or even hop due to pain within the ankle joint. This is usually due to synovitis (inflammation of the joint lining) of the ankle, with possibly some underlying damage to the joint surface. Specialized investigation and treatment are necessary.

(c) LEG INJURIES

Excessive running or overbowling are usually the reason for the development of pain in the front of the leg. This may be due to shin splints, stress fractures, tendonitis or at times a compartment syndrome (the latter is due to an increase in pressure in the compartment of the leg). All necessitate accurate diagnosis and appropriate treatment. None of these conditions is specific to cricket.

Achilles tendonitis is a common and troublesome condition which can afflict anyone who runs a lot. The covering of the tendon sheath or the tendon itself, or both, become inflamed. The result is pain on running, tenderness and slight swelling of and around the tendon, and stiffness, especially in the morning. The condition may remain mild and permit continued bowling assisted by physiotherapy, a heel lift and stretching exercises. It may, however, progress and prevent running. Occasionally in older bowlers who are 30 years of age or older, the Achilles tendon may rupture and this necessitates surgery, after which full recovery is still possible.

(d) KNEE INJURIES

While the knee remains the most complicated and difficult region of the body in the field of sports medicine, fast bowlers are not prone to any specific knee injury. They are, however, liable to sustain the same kinds of injuries that afflict many others.

Suddenly changing direction or decelerating as in trying to field or catch a ball that has been straight driven may damage the ligaments or menisci (cartilages) in the knee of a fast bowler. Adolescent fast bowlers, like many of their contemporaries engaged in other sports, are prone to problems related to the patellofemoral joint (joint involving the knee cap). Injuries previously sustained during sports participation can be aggravated by fast bowling. All knee injuries warrant accurate diagnosis and appropriate treatment. Unless rectified, a knee injury will reduce the effectiveness of a fast bowler.

(e) THIGH INJURIES

Muscle strains of the hamstring (back of thigh) and quadriceps (front of thigh) are not particularly common while bowling but may be sustained during training, batting or fielding due possibly to the requirement for sudden acceleration in different directions. Physiotherapy, improved stretching techniques and cessation of bowling until fully recovered are indicated.

3. Trunk Injuries

One condition which is almost exclusively encountered in fast bowlers is the 'rib tip syndrome'. This is a muscle strain affecting usually the 'tenth or eleventh' ribs close to the end of the rib. It is usually caused by attempting to bowl too fast, especially if not sufficiently warmed up, or bowling a 'bouncer'. In some instances, fatigue can also be a factor. The bowler experiences sudden pain on the non-bowling side in the vicinity of the lower ribs a little behind the nipple line. Further attempts to bowl are prevented because of the pain arising in this region during 'wind-up'. Immediate application of ice should follow and physiotherapy may be helpful. The condition is prone to be rather slow to heal and frequently benefits from an injection or two of a corticosteroid preparation to accelerate recovery. It is mandatory that bowling does not recommence for seven to ten days after the last injection.

4. Upper-limb Injuries

(a) SHOULDER INJURIES

Considering the stresses to which the shoulder appears to be subjected while fast bowling, it is surprising how infrequently this region is actually injured in the act of bowling. Injuries sustained during falls, while throwing, or occasionally at weight training, are the usual reasons why the shoulders of fast bowlers are injured. In all circumstances an accurate diagnosis is essential and the appropriate treatment instituted.

An unstable shoulder, either due to a past dislocation and repeated subluxation (partial or incipient dislocation), is probably the most potentially disabling shoulder problem. This invariably prevents classical throwing, though an 'underarm throw' is usually possible. Treatment is difficult and often necessitates surgery, which is why the occasional sufferer simply resorts to a substitute throwing action.

Tendonitis is a common shoulder problem and one that may prevent fast bowling. This may result from either excessive bowling or an incorrect action but is far more often a consequence of throwing. Rest from throwing and, if necessary, from bowling may be required. Physiotherapy and suitable stretching exercises are recommended as part of rehabilitation.

(b) ELBOW INJURIES

The elbow, like the shoulder, is subjected to much greater stresses during the action of throwing than when bowling. Inflammation of the joint (synovitis) or the creation of loose bodies, usually bony, within the joint do occur and cause an inability to straighten the elbow. Bowling under such circumstances can be both painful and classed by umpires as 'throwing' because the elbow is kept bent to reduce the pain. Treatment is difficult and necessitates advice from a suitable specialist.

CHAPTER 8

Rehabilitation of the Injured Bowler

DAVID JOHN, BRIAN BLANKSBY and SANDY GORDON

Previous chapters have explained how, despite carefully planned progressive training programmes, the explosive nature of fast bowling places stresses and strains on the body and creates a likelihood of injury. When an injury occurs it is essential that fast bowlers adopt proper rehabilitation procedures. These involve both physical and psychological components.

1. Physical Rehabilitation

The effectiveness of the rehabilitation programme determines the period the bowler is kept out of cricket. The risk of injury recurrence is also more likely to occur with players who do not follow a comprehensive rehabilitation programme.

The rehabilitation programme begins immediately after an injury, and finishes when the fast bowler recommences bowling without any limitations. The psychological aspects of the bowler's response to, and rehabilitation from, injury are discussed later in this chapter and need to be integrated within the three phases of physical rehabilitation, which are:

(a) initial treatment and diagnosis of the injury;
(b) rehabilitation;
(c) maintenance.

Special emphasis is given to remedial programmes for back injuries, because of the debilitating nature and prevalence of such injuries in fast bowling.

(a) INITIAL TREATMENT AND DIAGNOSIS PHASE

Sprains, strains and fractures are the most common injuries sustained by fast bowlers (Chapter 7). Each of these injuries usually causes damage to primary structures (tendons, ligaments, muscle or bone) and disruption to associated soft tissue structures (nerves, capillaries and blood vessels). The severity of an injury can often be determined by the degree of pain and swelling. The initial pain from an injury is the result of acute damage to nerve endings within the injured soft tissue, while local swelling is caused by disruption to blood vessels.

Therefore, initial treatment of an injury must aim to reduce the initial pain and control the amount of swelling to prevent secondary damage to the injured area.

The first priority following injury is **rest** to prevent further impairment to less severely damaged blood vessels and nerve endings. This refers to stopping the use of the injured area even to the extent of using crutches or some form of immobilization such as a splint, cast or brace.

The second priority is to **ice** the injured area. Ice curtails bleeding around the injury, which in turn decreases the amount of swelling, inflammation and pain. Though there are items such as chemical cold packs and cold whirlpools to cool the injured area, immersion in an ice bath or application of a wet cloth filled with ice are two effective techniques. Treatments for 20 to 30 minutes every 1–2 hours are desirable for the first 24–36 hours following an injury.

The third priority is **compression** of the injured area. The use of a compression bandage immediately following injury restricts the accumulation of fluid within the injured joint or muscle. Compression is particularly useful following an ice treatment. When ice is removed from the injured area, large quantities of blood are redirected to this section of the body to warm the area and, consequently, further internal bleeding may occur. However, the application of direct pressure to the injured area forces blood and serous fluid into the circulatory systems, thus controlling the degree of swelling.

The fourth priority in the early treatment of an injury is **elevation** of the injured area. The purpose of elevation is to allow gravity to decrease the accumulation of fluid around the injury. It is therefore important that the injured area is raised above the level of the heart to initially prevent swelling and then to promote fluid drainage from the area.

The final priority in the initial management of an injury is **diagnosis**. Any injury sustained by a fast bowler should be examined as soon as possible by a suitably qualified person, to determine the nature and extent of the injury. This is especially important if the back is injured, because the early diagnosis and treatment of back injuries can minimize the period away from the game and lower the risk of permanent damage.

(b) REHABILITATION

When an injury occurs, rest and/or immobilization of the injured body part is necessary for healing to take place. However, prolonged immobilization of an injured joint or muscle may result in decreased joint flexibility, articular cartilage degeneration, loss of connective and muscle tissue density, decreased muscle strength, deterioration of neuromuscular function and loss of aerobic fitness. Physical activity, a primary stimulus for growth and repair, may accelerate the healing process. Exercise should therefore be part of the rehabilitation process as soon as is medically advisable, to assist healing and to decrease the effects of the immobilization. It must be stressed, however, that the activity prescribed is initially of a mild nature and not at a level that would hinder the healing process or cause pain.

Rehabilitation is the process of returning an injured athlete to pre-injury levels of flexibility, strength, endurance, bowling technique, mental attitude and agility before returning to competition. Programmes should be designed and administered to ensure that flexibility, muscle strength, muscular endurance, aerobic fitness, anaerobic fitness, neuromuscular co-ordination, technique and a correct mental approach to the injury are all catered for in the sometimes complex preparation for return to the competitive environment.

Reconditioning generally requires that emphasis be placed on one facet of rehabilitation prior to moving to the next area. However, if an injury such as a sprained wrist does not prevent general cardiovascular conditioning then this aspect of training should be continued during the rehabilitation process. A discussion of each area in the generally accepted sequence of reconditioning, complete with sample exercises, follows to illustrate the rehabilitation of a lower back injury. A similar protocol could be followed for a variety of injuries sustained from fast bowling.

Flexibility.   Joint flexibility is generally reduced following injury and is due to either muscle spasm, pain, neural inhibition or contraction of connective tissues. It is very important to initiate flexibility exercises early in the rehabilitation process to ensure that the full range of joint motion is restored before strengthening exercises begin. However, the exercises should not cause pain. A player should stretch to the point of discomfort but not pain. The effectiveness of flexibility exercises may be facilitated by the prior application of either cold packs during the early stages of treatment, or hot packs when swelling and bruising have subsided. The exercises must be specific to the injured area which, in this example, is the lower back.

The following series of exercises are listed in increasing order of difficulty and serve to stretch and mobilize the lower back and abdominal regions (Table 8.1). Progression through the series will depend on the degree of pain associated with each exercise. That is, if there is no pain, continue with the next exercise.

TABLE 8.1

STRETCHING AND MOBILITY
EXERCISES

STRETCHING EXERCISES

Exercise 1

- ► Prone lying (face down), arms bent and close to sides, palms down just under the shoulders.
- ► Straighten the arms and raise the head and trunk to arch the back.
- ► Hold for 10 s, then gradually lower back to the starting position.
- ► Repeat three times with a 10 s rest between stretches.

Exercise 2

- ► Sitting on a chair with knees bent, feet flat on floor.
- ► Lean forward from the hips, to place the forehead between the knees.
- ► Hold at the limit of the movement for 10 s, then return gradually to the start position.
- ► Repeat three times with a 10 s rest between stretches.

Exercise 3

- ► Lying supine (on back) with knees bent.
- ► Place the right ankle over the left knee and use right leg to pull left leg to the ground.
- ► Keep shoulders in contact with the ground.
- ► Hold position for 10 s, then stretch in the opposite direction by pulling the right leg to the ground with the left foot.
- ► Repeat twice for both sides.

MOBILITY EXERCISES

Exercise 1

- ► Lying supine, knees bent, feet flat on floor.
- ► Alternately flatten the lumbar spine onto the floor, then arch, keeping hips and shoulders in contact with the floor.
- ► Repeat a set of up to ten repetitions in succession.
- ► Repeat three times with 60 s rest between each set.

Exercise 2

► Lying supine, knees bent, feet flat on floor.
► Keeping feet and shoulders in contact with the floor, push hips up into the air, then lower to the floor.
► Repeat this movement up to ten times in succession.
► Repeat three times with 60 s rest between each set.

Exercise 3

► Lying supine, knees bent and together, feet flat on floor.
► Roll knees alternately to touch the floor on each side of the body.
► Keep both shoulders in contact with floor.
► Repeat up to ten times on each side.

Exercise 4

► Lying supine.
► Raise knees to chest, roll both knees to each side to touch the floor.
► Keep both shoulders in contact with floor at all times.
► Repeat up to ten times on each side.

Exercise 5

► Kneeling on hands and knees.
► Keeping the arms and legs stationary, arch the back upwards while looking downwards, then hollow back while raising head upwards.
► Repeat movement up to ten times.

Strength. When sufficient flexibility has been gained in the injured area such that a full range of movement can be completed without pain, specific strengthening exercises should be introduced to minimize muscle atrophy, and provide a protective and supportive mechanism for the injured area. In the case of a back injury, muscle strength of both the anterior abdominal muscles and the posterior back extensor muscles should be increased. In fact, a deficiency of strength in either of these two regions may have been one of the primary causes of the injury in the first place. Research has suggested that poor abdominal and lower back strength may be a contributing factor in back injuries to fast bowlers.

Progressive resistance exercises must be performed regularly to improve strength to at least pre-injury levels, because muscle atrophy occurs very quickly. When an injury occurs, knowledge of pre-injury levels provides accurate information as to the strength level required and to which the player must aspire during rehabilitation. However, efforts to strengthen injured muscles too quickly may cause inflammation and delay the healing process. Therefore, strength programmes should provide graduated exercises to regain maximal strength levels in the injured muscles in a safe manner.

Two different programmes that aim to strengthen the abdominal (Table 8.2) and back extensor (Table 8.3) muscle groups are provided. Before beginning a strength training programme it is essential that none of the exercises to be attempted will aggravate the current injury. For example, lumbar flexion exercises (sit-ups) may aggravate a displaced or herniated disc.

TABLE 8.2

ABDOMINAL STRENGTHENING EXERCISES

SIMPLE EXERCISES (2–3 WEEKS)

Exercise 1: Head raising

► Lying supine, with hands resting on thighs and legs straight.

► Lift the head toward the knees, by sliding the hands down the thighs.

► Raise the head only until both shoulder blades are clear of the floor.

► Hold for 6 s, then gently lower.

Week 1 — 10 repetitions
Week 2 — 15 repetitions
Week 3 — 20 repetitions

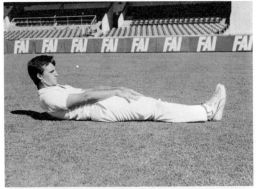

Exercise 2: Lower-limb raising

► Lying supine with one limb flexed to 90°, and the other limb straight.

► Keeping the knee of the straight limb tensed, raise *slowly* to the level of the bent knee.

► Lower *slowly* to the floor.

Week 1—15 repetitions with each leg
Week 2—20 repetitions with each leg
Week 3—25–30 repetitions with each leg

MODERATE EXERCISES (2–3 WEEKS)

Exercise 1: The curl

► Lying supine with the knees flexed to 90°, and the feet flat on the floor.

► Lift both knees toward the chest and then lower slowly.

► Keep knees bent at all times.

Week 1—15 repetitions
Week 2—20 repetitions
Week 3—30 repetitions

Exercise 2: Crunches

► Lying supine with the knees flexed, hands on the forehead.

► Lift both knees and bring the head and shoulders forward towards the knees.

► Slowly return to starting position.

Week 1—15 repetitions
Week 2—20 repetitions
Week 3—30 repetitions

Exercise 3: Sit-ups

► Lying supine with the knees flexed to 90°, feet flat on the floor and hands on forehead.

► Lift the head and shoulders to touch the forehead to the knees (feet may need to be fixed).

► Slowly return to the start position.

Week 1—15 repetitions
Week 2—20 repetitions
Week 3—30 repetitions

Exercise 4: Sit-ups with a twist

► Lying supine with the knees flexed to 90°, hands on the forehead and feet flat on floor.

► Lift the head and shoulders to touch the left elbow to the right knee and then return to starting position.

► Repeat, touching the right elbow to the left knee.

Week 1—15 repetitions
Week 2—20 repetitions
Week 3—30 repetitions

ADVANCED EXERCISES

Exercise 1: Thigh raise

► Lying supine with the knees bent to 90°.

► Tighten the abdominal muscles.

► Lift both knees towards the chest and then lower the legs to the start position.

Week 1—15 repetitions
Week 2—20 repetitions
Week 3—30 repetitions

Exercise 2: Gluteal flexor

► Lying supine with the knees flexed to 90°.

► Place the hands on the abdominal region, raise both knees and then straighten the right limb with the foot flexed. Return the right limb to flexed knee position.

► Repeat with the left limb outstretched.

Week 1—15 repetitions with each leg
Week 2—20 repetitions with each leg
Week 3—30 repetitions with each leg

Exercise 3: Oblique raise

► Lying supine with the legs straight.

► Bend the left leg to 90°.

► Let the left knee drop slowly to the left and then tighten the abdominal muscles to help lift the knee back to the vertical.

► Repeat for both sides.

Week 1—10 repetitions each side
Week 2—15 repetitions each side
Week 3—20 repetitions each side

Exercise 4: Heel squeeze

► Lying supine with the legs straight.

► Squeeze the heels together, with the feet flexed.

► Raise the legs as high as comfortably possible while able to keep the legs straight.

► Tighten the abdominal muscles and let the limbs drop 3–4 cm to the right and then left. The heels can also be circled in the air.

Week 1—15 repetitions
Week 2—20 repetitions
Week 3—25 repetitions

TABLE 8.3

BACK EXTENSOR STRENGTHENING EXERCISES

Exercise 1: Prone spinal extension

► Lying prone with the hands placed beneath the shoulders. Exhale and straighten arms as if doing a push-up. Keep hips flat on floor and hold the position for 10 s.

► Return to the starting position, relax for 10 s and repeat 20 repetitions.

Exercise 2: Leg-overs

► Back lying with legs fully extended and arms extended to the sides. Keeping the right leg straight, roll the foot over to try and touch the ground on the left side of the body. Hold for 5 s while keeping both shoulders in contact with the floor.

► Return to starting position and roll the left leg over to the right hand. Perform 10 repetitions to each side.

Exercise 3: Lower-limb extension

► Prone lying with the pelvis pressed onto the floor.

► Raise one limb approximately 30 cm and lower to floor.

► Repeat using the other limb.

► Complete 20 repetitions for each limb.

Muscular endurance. Following an injury, immobilization or rest will probably reduce muscular endurance or the ability to sustain repeated contractions over a long period. By using increased repetitions of lighter weights in preference to heavier loads, muscular endurance is regained in parallel with the development of some muscle strength.

One would normally choose a load or weight that could be handled relatively easily for at least ten to fifteen repetitions in a set at a comfortable speed. Delayed muscle soreness may indicate an overload, in which case the frequency and/or the intensity of the exercises should be reduced.

Aerobic fitness. Rest or immobilization of an injury is necessary to facilitate the healing process. It is necessary, however, that the athlete participates in some form of exercise as soon as possible during the rehabilitation phase to limit the loss of aerobic fitness. Studies have shown that cardiovascular fitness can begin to decrease after one week of no training and can return a person to sedentary levels after eight weeks of inactivity. A player should continue in some form of aerobic exercise(s) which will not aggravate the injury during the recovery process. For example, an injured back or leg may contra-indicate running, but swimming could be valuable. An injured arm or shoulder does not prevent one from running, or possibly cycling, and maintaining strength levels of both uninjured lower limbs and the uninjured upper limb.

Different types of cardiovascular exercise such as swimming, cycling, pool running, brisk walking or running are valuable for an injured fast bowler to help maintain or improve aerobic fitness and to condition the muscles used in the bowling action.

A typical six-week training programme of swimming and pool running for a fast bowler with an injured back is outlined in Table 8.4. However, the type and intensity of each workout should be specific to the bowler's level of fitness and skill, and must not aggravate the current injury.

TABLE 8.4

SWIMMING AND POOL RUNNING PROGRAMMES

Swimming programme

Note that all work is done gently with no jerky movements of the arms, head or legs.

Week 1

► WARM-UP: 6 x 25 m freestyle at a comfortable pace (15 s rest interval between each 25 m).

► STRETCH: Stretch the arm, shoulder, back and lower-limb muscles as shown in Chapter 2.

► WORKOUT: (a) 4 x 25 m freestyle using arms only with pull buoy.
    Rest: 15 s between each 25 m.

        (b) 4 x 25 m gentle patter kick on front with face in the water. Use the breast-stroke arm action and turn head to the side when a breath is needed.
    Rest: 60 s between each 25 m of kicking.

        (c) 2 x 25 m catch-up arms with pull buoy.
    Rest: 45–60 s between each 25 m.

        (d) 2 x 25 m patter kick on back.
    Rest: 45–60 s between each 25 m.

        (e) 6 x 25 m freestyle swims with moderate effort.
    Rest: 30 s between each 25 m.

► COOL-DOWN: 4 x 25 m easy pace with pull buoy, 10 s rest between each 25 m.

TOTAL 550 m

Week 2

▶ WARM-UP and STRETCH: As for Week 1.

▶ WORKOUT: (a) 2 x 25 m patter kick on front with face in water.
Rest: 45–60 s between each 25 m.

        (b) 6 x 25 m freestyle using arms with pull buoy at moderate pace.
Rest: 30 s between each 25 m.

        (c) 1 x 50 m back kicking.
Rest: 60 s.

        (d) 2 x 25 m patter kick on front (as (a)).
Rest: 45–60 s between each 25 m.

        (e) 2 x 50 m freestyle swim at steady pace.
Rest: 60 s between each 50 m.

        (f) 2 x 50 m at 75% pace.
Rest: 60 s between each 50 m.

        (g) 2 x 25 m at 100% pace.
Rest: 60 s between each 25 m.

▶ COOL-DOWN: 100 m freestyle swim at easy pace with pull buoy.

TOTAL 650 m

Week 3

▶ WARM-UP and STRETCH: As for Week 1.

▶ WORKOUT: (a) 4 x 50 m freestyle with pull buoy.
Rest: 45 s between each 50 m.

        (b) 1 x 50 m patter kick on front with face in water.
Rest: 60 s.

        (c) 1 x 50 m patter kick on back.
Rest: 90 s.

        (d) 2 x 25 m catch-up arms with pull buoy.
Rest: 45 s between each 25 m.

        (e) 1 x 100 m freestyle swim at 50% max. pace.
Rest: 60 s.

        (f) 4 x 50 m at 75% max. pace.
Rest: 60 s between each 50 m.

        (g) 4 x 25 m effort swims.
Rest: 60 s between each 25 m.

▶ COOL-DOWN: 1 x 100 m gentle swim (freestyle or backstroke).

TOTAL 850 m

Week 4

▶ WARM-UP: 1 x 100 m freestyle at 50% max.

▶ STRETCH: As for Week 1.

▶ WORKOUT: (a) 4 x 50 m freestyle pull with pull buoy.
                   Rest: 60 s between each 50 m.

                (b) 2 x 25 m backstroke kick with kickboard on chest.
                   Rest: 45 s between each 25 m.

                (c) 2 x 25 m easy backstroke swim.
                   Rest: 30 s between each 25 m.

                (d) 2 x 50 m arms only with pull buoy between legs.
                   Rest: 60 s between each 50 m.

                (e) 2 x 100 m freestyle swim at easy pace.
                   Rest: 60 s between each 100 m.

                (f) 4 x 50 m at 75% max. pace.
                   Rest: 60 s between each 50 m.

                (g) 4 x 25 m at 90% pace.
                   Rest: 60 s between each 25 m.

▶ COOL-DOWN: 1 x 100 m at easy pace.

TOTAL 1100 m

Week 5

▶ WARM-UP: 1 x 100 m easy freestyle swim

▶ STRETCH: As for Week 1.

▶ WORKOUT: (a) 4 x 50 m with pull buoy.
                   Rest: 30–45 s between each 50 m.

                (b) 6 x 25 m patter kick on front with face in water.
                   Rest: 45 s between each 25 m.

                (c) 4 x 25 m backstroke swim at easy pace.
                   Rest: 15 s between each 25 m.

                (d) 2 x 25 m backstroke arms only with pull buoy.
                   Rest: 30 s between each 25 m.

                (e) 2 x 100 m freestyle swim at easy pace.
                   Rest: 60 s between each 100 m.

                (f) 6 x 50 m at 75% max. pace.
                   Rest: 60 s between each 50 m.

                (g) 4 x 25 m at 100% max. pace.
                   Rest: 60 s between each 25 m.

▶ COOLDOWN: 1 x 100 m at 50% max. pace.

TOTAL 1200 m

Week 6

► WARM-UP: 1 x 100 m easy freestyle swim.

► STRETCH: As for Week 1.

► WORKOUT: (a) 2 x 100 m with pull buoy.
              Rest: 60 s between each 100 m.

           (b) 3 x 50 m patter kick on front, face in the water.
              Rest: 90 s between each 50 m.

           (c) 4 x 50 m arms, legs and breathing with pull buoy used as a kickboard.
              Rest: 60 s between each 50 m.

           (d) 2 x 25 m freestyle kick on the side (no arms).
              Rest: 30–45 s between each 25 m.

           (e) 3 x 100 m at 50% max. pace.
              Rest: 60–90 s between each 100 m.

           (f) 4 x 50 m at 75% max. pace.
              Rest: 60 s between each 50 m.

           (g) 6 x 25 m at 95% max. pace.
              Rest: 60 s between each 25 m.

► COOL-DOWN: 1 x 100 m freestyle at easy pace.

TOTAL 1350 m

POINTS TO NOTE:

• Pull buoy: A swimming aid which is placed between the thighs to assist buoyancy of the lower body and isolates the effort to the arms only.

• CATCH-UP ARMS: Swimming freestyle leaving one arm out in front until the other arm catches it up; then pulling with the rested arm.

• ARMS, LEGS AND BREATHING: Holding the pull buoy out in front, one swims the whole stroke with a 'catch up' action which forces the legs to work consistently.

• PATTER KICK: Propulsion using the lower limbs without the use of the arms which are placed out in front and under the water—not holding a board.

• If the player is a trained swimmer, the above repetitions and distances could be doubled and/or the rest intervals halved to optimize the training benefit.

POOL RUNNING PROGRAMME

Week 1

► STRETCH: Stretch the lower-back and lower-limb muscles as shown in Chapter 2.

► WORKOUT: (a) Walk slowly in knee- to waist-deep water for 90 s using a normal but vigorous arm action.
              Rest: 90 s.

           (b) Walking at a slightly faster pace in knee- to waist-deep water for 60 s using same arm action.
              Rest: 90 s.

   (c) Walking at a fast pace in knee- to waist-deep water for 30s using the same vigorous arm action.

       Rest: 90s.

   (d) Use an air vest or pull buoy under each armpit to suspend the body in the deep water. Perform an easy running action in deep water for 90s without using the arms.

       Rest: 90s.

   (e) Running at a slightly faster pace while suspended in deep water for 60s without using the arms.

       Rest: 90s.

   (f) Running as fast as possible without incurring pain while suspended in deep water for 30s without using the arms.

       Rest: 90s.

► COOL-DOWN: 1 x 100 m freestyle at an easy pace.

Week 2

► STRETCH: As for Week 1.

► WORKOUT: (a) Normal forward walking for 30s, side-stepping to each side for 30s in knee- to waist-deep water.

       Rest: 60s.

   (b) Walking at a slow pace in knee- to waist-deep water for 90s with a normal but vigorous arm action.

       Rest: 90s.

   (c) Walking at a slightly increased pace in knee- to waist-deep water for 90s with arm action.

       Rest: 90s.

   (d) Walking at maximum speed in knee- to waist-deep water for 45s with arm action.

       Rest: 90s.

       Using an air vest or pull buoys to suspend the body in deep water.

   (e) Running at an easy pace in deep water for 90s with no arm action.

       Rest: 90s.

   (f) Running at a slightly faster pace in deep water for 90s with no arm action.

       Rest: 90s.

   (g) Running as fast as possible without incurring pain in deep water for 2 x 30s with no arm action.

       Rest: 90s between each 30s repetition.

► COOL-DOWN: 1 x 100 m freestyle at easy pace.

Week 3

► STRETCH: As for Week 1.

► WORKOUT: (a) Forward walking (30s), side-stepping to each side (30s) and backward walking (30 s) in knee- to waist-deep water using a vigorous arm action. Rest: 90 s.

(b)   Walking at a slow pace in knee- to waist-deep water for 90 s with a normal but vigorous arm action.
Rest: 90 s.

(c)   Walking at an increased pace in knee- to waist-deep water for 2 x 45 s with the same arm action as above.
Rest: 90 s between each 45 s repetition.

(d)   Walking at maximum pace in knee- to waist-deep water for 2 x 30 s with a normal but vigorous arm action.
Rest: 90 s between each 30 s repetition.

Using an air vest or pull buoys to suspend the body in deep water.

(e)   Running at an easy pace in deep water for 4 x 45 s with no arm action.
Rest: 90 s between each 45 s repetition.

(f)   Running at a slightly faster pace in deep water for 4 x 45 s with no arm action.
Rest: 90 s between each 45 s repetition.

(g)   Running as fast as possible without incurring pain in deep water for 2 x 30 s with no arm action.
Rest: 90 s between each 30 s repetition.

► COOL-DOWN: 1 x 100 m freestyle at an easy pace.

Week 4

► STRETCH: As for Week 1.

► WORKOUT: (a)   Forward walking (30 s), side-stepping to each side for (30 s) each and backward walking (30 s) in knee- to waist-deep water using a vigorous arm action.
Rest: 90 s.

(b)   Walking slowly in knee- to waist-deep water for 3 x 45 s with normal but vigorous arm action.
Rest: 90 s between each 45 s repetition.

(c)   Walking at a slightly increased pace in knee- to waist-deep water for 3 x 45 s with a vigorous arm action.
Rest: 90 s between each 45 s repetition.

(d)   Walking as fast as possible in knee- to waist-deep water for 3 x 30 s with a vigorous arm action.
Rest: 90 s between each 30 s repetition.

Using an air vest or pull buoys to suspend the body in deep water.

(e)   Running at an easy pace in deep water for 4 x 45 s with no arm action.
Rest: 60 s between each 45 s repetition.

(f)   Running at a slightly faster pace in deep water for 4 x 45 s with no arm action.
Rest: 60 s between each 45 s repetition.

(g) Running as fast as possible without incurring pain in deep water for 3 x 30 s with no arm action.

Rest: 90 s between each 30 s repetition.

► COOL-DOWN: 1 x 100 m freestyle swim at easy pace.

Week 5

► STRETCH: As for Week 1.

► WORKOUT: (a) Walking slowly in knee- to waist-deep water for 4 x 45 s with a normal but vigorous arm action.

Rest: 60 s between each 45 s repetition.

(b) Walking at a steady pace in knee- to waist-deep water for 4 x 45 s with a vigorous arm action.

Rest: 90 s between each repetition.

(c) Walking at a fast pace in knee- to waist-deep water for 4 x 30 s with a vigorous arm action.

Rest: 90 s between each 30 s repetition.

Using an air vest or pull buoys to suspend the body in deep water.

(d) Running at an easy pace in deep water for 5 x 45 s with an arm action if a vest used.

Rest: 60 s between each 45 s repetition.

(e) Running at a slightly faster pace in deep water for 5 x 45 s with a normal arm action if a vest used.

Rest: 60 s between each 45 s repetition.

(f) Running as fast as possible without incurring pain in deep water for 5 x 30 s with a normal arm action.

Rest: 90 s between each 30 s repetition.

► COOL-DOWN: 1 x 100 m easy freestyle swim.

Week 6

► STRETCH: As for Week 1.

► WORKOUT: (a) Walking slowly in knee- to waist-deep water for 5 x 45 s with a normal but vigorous arm action.

Rest: 60 s between each 45 s repetition.

(b) Walking at a steady pace in knee- to waist-deep water for 5 x 45 s with a normal but vigorous arm action.

Rest: 90 s between each 45 s repetition.

(c) Walking at a fast pace in knee- to waist-deep water for 5 x 30 s with a normal but vigorous arm action.

Rest: 90 s between each 30 s repetition.

Using an air vest or pull buoys to suspend the body in deep water.

(d) Running at an easy pace in deep water for 6 x 45 s with a normal arm action if a vest is used.

Rest: 60 s between each 45 s repetition.

(e) Running at a slightly faster pace in deep water for 6 x 45 s with a normal arm action.

Rest: 60 s between each 45 s repetition.

(f) Running as fast as possible without incurring pain in deep water for 6 x 30 s with a normal arm action if using a vest.

Rest: 90 s between each 30 s repetition.

► COOLDOWN: 1 x 100 m freestyle swim at an easy pace.

As a final note, it is worthwhile observing that if exercises of these levels of frequency and intensity are necessary for rehabilitation, the same levels are needed for maintaining fitness. An injured bowler who finds the rehabilitation programme more strenuous than the normal activities involved in playing and training may well have under-preparation to blame for the injury.

Anaerobic fitness.   Speed is a component of the anaerobic system and must be redeveloped when the player has regained at least pre-injury levels of flexibility, muscle strength and endurance, and aerobic fitness. It is very important that an injured muscle or group of muscles can function at maximum speed before the athlete returns to competition. Muscle contraction speed should be developed through sport-specific exercises and drills. Exercises and drills for fast bowlers which help to develop muscle speed in the lower back and lower limbs are presented in Table 8.5. While an emphasis has been placed on explosive-type activities because of the dynamic nature of fast bowling, these activities must always be preceded by flexibility and general warm-up procedures.

There are a number of ways to improve lower limb power and speed, but essentially the aim is to:

- increase the strength of the limbs;
- increase the rate of limb movement.

Common methods of achieving some or all of these aims are shown in Table 8.5.

TABLE 8.5

AN ANAEROBIC REHABILITATION PROGRAMME

PLYOMETRIC DRILLS

e.g.   • Performing vertical jumps on the spot for 30–60 s at maximum speed using a double-foot take-off and double-arm swing.

- Hopping over distances of 20–50 m using long and forceful efforts. This can be varied by using alternate limbs and single-arm or double-arm action.

- Double-leg take-offs to land on one leg over a series of small hurdles (< 0.5 m) placed 2 m apart.

- Jumping downward from a height (30–60 cm) and exploding upwards to a similar height.

SPRINT STARTS

e.g.   • Assume a crouched start position (starting blocks if possible), explode from the blocks and sprint for 10–30 m.

• Alternate lower limb positions to develop power in each limb.

UPHILL AND DOWNHILL RUNNING

e.g.   • Complete 15–20 repetitions over short distances of 20–50 m at fast speeds on mild, uphill gradients to achieve increased knee lift and leg strength.

• Complete 5–10 repetitions over longer distances of 50–100 m at fast speed on 5% downhill gradients to achieve increased stride rate.

Neuromuscular co-ordination. Prolonged immobilization may also result in reduced neuromuscular co-ordination and proprioception in a joint and the surrounding muscles, tendons and ligaments. Research studies have shown that decreased muscle control and proprioception can be a primary cause of re-injury. Therefore, it is important to restore co-ordination to pre-injury levels before competitive bowling is resumed. Practising sport-specific movements and various kinaesthetic exercises can redevelop neuromuscular co-ordination and enhance proprioceptive function. For example, wobble boards are commonly used to develop kinaesthesis following an ankle injury (Figure 8.1).

Figure 8.1: Wobble board training

A fast bowler recovering from, say, a back injury should practise various components of the bowling action such as the run-up and delivery stride at sub-maximal pace to facilitate proper neuromuscular functioning before bowling at full pace and resuming competition.

(c) MAINTENANCE PHASE

The final stage of any return to full fitness should involve the continuation of exercises to maintain the levels of flexibility, muscle strength and endurance, aerobic and anaerobic fitness attained during the rehabilitation phase. Maintenance and further development of these physical characteristics may help to prevent re-injury once the athlete has returned to competition. Players should also be encouraged to have periodic fitness evaluations (every four weeks) to monitor general fitness and the condition of the rehabilitated area.

Running and weight-training activities are recommended for fast bowlers in their maintenance programme. A typical six-week running programme for fast bowlers with a history of back injury is outlined in Table 8.6, while a modification of the general weight-training programme for beginners in Chapter 2 would enable the fast bowler to maintain strength and endurance.

TABLE 8.6
MAINTENANCE PHASE RUNNING PROGRAMME

Week 1
- ► Mon: 4–5 km   Slow pace (include some walk/slow run, with intermittent high knee lifts to prevent hyperextension of the back).
- ► Thu: 5–6 km   Slow pace (as above).
- ► Sat: 3–5 km   Medium pace (depends on soreness).

Week 2
- ► Mon: 6–7 km   Slow pace.
- ► Thu: 4–5 km   Medium pace.
- ► Sat: 3–4 km   75% of pace achieved for the 12-min. run test.

Week 3
- ► Mon: 6–7 km   Slow-medium pace.
- ► Thu: 5–6 km   Medium pace.
- ► Sat: 3–4 km   2–2.5 km fast pace; 3–5 repetitions of 30–50 m run-through efforts with 30–60 s walk in between each run; 0.5–1.0 km gentle jog to cool down.

Week 4
- ► Mon: 4–5 km   Fast pace.
- ► Thu: 5–6 km   3–4 km medium pace; 5–8 repetitions of 30–50 m run-through efforts with 30–60 s walk in between each run; 1 km gentle jog to cool down.
- ► Sat: 4 km   Fast pace.

Week 5
- ► Mon: 5 km   2–3 km medium pace; 15 x 30 m and 10 x 50 m, 90% effort sprints, work to rest ratio 1:5 (i.e. run for 5 s, rest for 25 s); 1–2 km gentle jog to cool down.

► Thu: 4 km   Medium pace.
► Sat: 4–5 km Light jog.

Week 6
► Perform a 12-min. run to gauge level of recovery, by comparing with previous times (norms for adult performance are included in Chapter 2).

2. Psychological Aspects of Response to and Rehabilitation from Injury

The physical consequences of athletic injuries can now be detected, diagnosed and treated by health care professionals such as physicians, sport physiotherapists and physical educators. The psychological consequences of athletic injuries, however, are neither easy to detect and diagnose, nor widely understood.

Recent attention in sport psychology on sport injuries has produced an important body of literature and some applied research evidence. Professionals in both the medical and sport communities, and all athletes, should be acquainted with this research.

(a) PSYCHO-SOCIAL CAUSES OF INJURY

Cricket injuries, like most sport injuries, are often regarded as 'occupational hazards' or just part of the game. But injuries do not just happen, they are caused.

Some injuries to bowlers are relatively 'uncontrollable'. They are often regarded as 'accidents' and may be caused by the player's interaction with the sport environment such as slipping on a wet surface, or turning an ankle in a rut on the crease; or interaction with other persons and objects, such as colliding with other players or being hit by the ball. The most common type of injury that occurs, however, can be described as self-induced, because the injury is often caused by improper technique, lack of adequate warm-up,or physical preparation, or overuse. Self-induced injuries are generally 'controllable' in that coaches, trainers and bowlers can minimize their occurrence by undertaking a balanced exercise programme and having a responsible attitude to training. Additional causes of self-induced injury may be related to certain psychological factors that can predispose some bowlers to both injury and re-injury. Sport studies have shown that highly anxious and worried athletes, and those who are described as over-confident, are often tagged 'injury-prone'. Similarly, 'tough' athletes who choose to play through pain and fatigue in order to 'prove something' are at risk of injury, as are all bowlers who have been confronted with significant life change events (e.g. death of a family member, spouse or close friend; marriage breakdown or other personal problems; career change or difficulties; move to a new city, house, team or coach).

The results from research that has examined a causal relationship between certain psychological phenomena and injury are at best equivocal, and so personality factors and life change events should not be used in any predictive sense. More research is required and particularly so in fast bowling where the most serious cricketing injuries often occur. In the meantime, attention to predisposing physical and psychological factors should be a priority concern of cricket administrators, coaches and bowlers in order to provide a safe environment for athletic development and the pursuit of excellence.

(b) RESPONSE TO INJURY

Bowlers' reactions to an injury are critical and each response is a highly subjective experience. However,the emotions suffered usually parallel those exhibited in a grief response following the death of a loved one. For example, a fast bowler with a serious back injury may react in the following way:

DENIAL: 'I'm OK, it's really not as bad as it seems.'

ANGER: 'Why me!? Why now!?—it's not fair!'

BARGAINING: 'OK I'm hurt—I'll see the specialist but only when it suits me. What do they know? I know what's best for me.'

DEPRESSION: 'I'm so sore, I can't do anything, it's hopeless—it's over.'

ACCEPTANCE: 'What's happened has happened—no point worrying about that anymore is there? Now what am I going to do about it . . .? What do I have to do to help myself?'

Whether or not these stages of response are actually sequential, progressive, or even evident, is not as critical for bowlers, coaches and trainers as it is to ensure that injured bowlers achieve the transition from 'denial' to 'acceptance' as quickly as possible. Bowlers who do not fully accept their injury will clearly not be able to concentrate totally on the rehabilitation process and may consequently jeopardize the likelihood of a successful recovery.

Health care professionals should be aware that private anxiety and grief are rarely exhibited by injured athletes. In general, injured athletes seldom appear publicly distressed and prefer instead to grieve in private. A 'macho' outward demeanour, cheerful resignation or other elaborate psychological defences can fool even the experienced eye of physicians and close friends. The important point is that one should not automatically assume that injured bowlers are coping psychologically with the injury. They actually should be asked how they feel about the injury and the response carefully evaluated.

Eight major stages or adaptive responses have been identified to effectively cope with the physical and emotional pain that accompanies serious injury. For fast bowlers these eight stages are as follows:

- Respond positively to the incapacitation and pain from the injury by informing the coach and medical staff immediately. Denial is plainly counterproductive and could result in a premature end to a cricket career.

- Adapt promptly to the stresses of treatment procedures. Visits to hospitals and clinics, and exposure to certain machines and orthopaedic devices often have an unsettling effect at first. For a normally healthy and active individual this 'adaptive response' will require perseverance.

- The player must have trust in the medical personnel and communication is essential. Victims of personal tragedies often have difficulty in coping with their sudden vulnerability and dependence on others. Diminished trust in professionals in unfamiliar settings is quite typical, but improvements are possible with just a little effort.

- Maintain an emotional balance. Injured athletes sometimes experience guilt such as 'I was told to warm-up before doing that' or, 'I was told to wait a few weeks before trying that again', or 'I've let the team down'. Bowlers can learn from mistakes and errors in judgement but should not brood on them. They should forgive themselves and get on with life and what they have to do for themselves.
- Maintain a healthy self-image. Because a sense of mastery and competence is suddenly lost, the injured bowler often loses confidence, particularly when a future in cricket is in the balance. Cricketers may need to re-evaluate career goals and ambitions and re-channel interests in areas other than cricket. This is often more difficult to do in reality than it would seem. Communication with former cricketers, friends, family and others whose opinions are respected can help.
- Preserve relationships with team mates, coaches, family and friends. Feelings of isolation, or even antagonism, may alienate injured bowlers from social support systems that they can depend on in troubled times. In order not to disrupt such relationships, the injured bowler must make the effort to maintain these associations. One should visit training sessions and games, and participate in normal social activities as often as possible. Everyone will be glad and relieved to see an injured bowler during the various recovery stages.
- Look forward positively to an uncertain future. Advances in surgery and other medical treatments can raise hopes of full recovery to active participation from quite serious injuries. The injured bowler should always be enthusiastic and positive about full recovery based on expert orthopaedic opinion and procedures. Injured bowlers who refuse surgery as an option or ignore expert opinion must consider whether or not they are acting in their own best interests.
- Accept the limitations and restrictions imposed by the injury and adjust lifestyle accordingly. Accept what **cannot** be done anymore and focus on what **can** be done. This strategy characterizes proud and brave disabled athletes who provide excellent models for prematurely retired athletes. This strategy is also appropriate for injured bowlers who may have to adapt their style and technique in order to continue playing.

As the foregoing suggests, injured bowlers can be taught to develop mental strategies that confront and combat pain, as well as the external pressures and various inconveniences associated with injury. Certain strategies are also particularly useful during rehabilitation, which is often as stressful as the occurrence of the injury itself.

(c) PSYCHOLOGICAL ASPECTS OF THE REHABILITATION PROCESS

The vast majority of injured cricketers probably return to play quickly and safely with just physical rehabilitation. However, others are less fortunate and do not return as quickly or as safely due to the extent and seriousness of their injuries. Dealing with constant pain and disablement is often a new and disturbing experience and so too is the isolation from friends, various inconveniences associated with treatment and, most of all, hours of compulsory inactivity. The injured cricketer must learn to cope with the psychological impact of these factors because full recovery depends upon both psychological and physical performance during rehabilitation.

A COGNITIVE BEHAVIOUR MODIFICATION MODEL OF HYPOTHESISED RESPONSES TO ATHLETIC INJURY

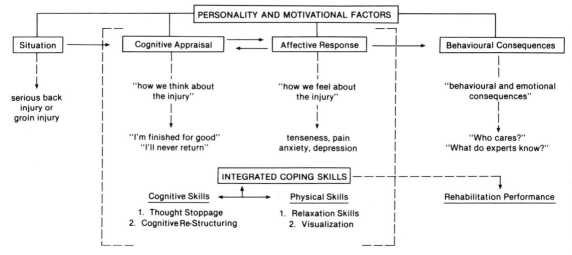

Figure 8.2: A behaviour modification model

Cognitive behaviour modification.   Cricketers who perform poorly during rehabilitation very often have an 'attitude problem'. They may find all sorts of excuses for not attending clinic sessions or for not doing exercises and individual training at home (behavioural problem). They may also develop elaborate self-defeating thought processes to rationalize this behaviour and their non-adherence to or non-compliance with rehabilitation protocols (cognitive problem). Cognitive behaviour modification (CBM) techniques are specifically designed to effect self-control over dysfunctional behaviour by modifying appraisal processes and inner experiences of the individual. A CBM model (Figure 8.2) illustrates how thoughts, feelings, physiological and emotional experiences which influence each other can influence behaviour. In terms of effective rehabilitation, therefore, the actual extent of the cricketer's injury may not be as critical as how he or she evaluates and feels about the injury.

Examples of using the 'thought stoppage ' technique are given in Table 8.7 to illustrate how dysfunctional and counterproductive self-talk can be prevented from having a negative effect on rehabilitation performance.

TABLE 8.7
EXAMPLES OF THOUGHT STOPPAGE USE

Stress situation	Self-defeating inner-dialogue	Self-enhancing modified inner-dialogue
1. 7.30 a.m. The clinic is closed and the trainer fails to appear for a regular treatment session.	'That idiot! Gives me the lecture about the importance of regular treatments in rehabilitation and then doesn't show up! She'd better apologize before I show up again. I've had it!'	'I'm really mad—STOP! But this is strange. Something important must have delayed her because I know how punctual and professional she is and how much she wants me to recover. I must get treatment regularly. I'll check back later to see if she can schedule another time for me.'
	Consequence: Athlete assumes no responsibility and fails to realize who is suffering most by not receiving therapy.	**Consequence:** Athlete rationalizes non-appearance of the therapist. Takes on responsibility for personal efforts to recover.
2. Athlete experiences great pain and discomfort during exercises and notices no significant improvements to the injury.	'This really hurts! I'm sick and tired of the pain! And there's no improvement! I need a break—they don't give a hoot—and if it really is important they'll call me. This will heal itself in time anyway.'	'Boy this hurts a lot—STOP! But this pain is what I was told to expect. If it gets too bad I guess I'd better tell them because they'll want to know. But I must persevere and try to get through this pain. That way I'll soon get out of here and be playing again.'
	Consequence: Athlete benefits little from the token effort expended during treatment. Develops and rationalizes excuses for goofing off.	**Consequence:** Athlete gives 100% in a quality session and prepares for more of the same; expresses difficulties and concerns to the therapist who positively encourages him and appreciates his attitude; feels good for the rest of the day because his behaviour is becoming more helpful and less inclined to self-pity.

Relaxation skills can help injured cricketers cope with the stresses of treatment. A relaxed state also facilitates healing and therefore may accelerate recovery from injury. Three relaxation techniques that could be employed during rehabilitation are deep breathing exercises, progressive muscular relaxation (PMR) and autogenic training:

- Deep breathing exercises assist cricketers to control arousal levels prior to and during treatment. Concentrating on the internal sensations created by deep diaphragmatic breathing internalizes the cricketer's focus of attention (and away from fearful events) and assists in conserving the vital energy required to promote healing.

- PMR teaches cricketers to recognize the increase and decrease of tension in different muscle groups. This technique is particularly useful when preparing for treatment and for dealing with localized pain experienced during treatment.

- Autogenic training works through self-suggestion and focuses on both physical relaxation (sensations of warmth and heaviness) and mental relaxation (visualization). In medical and sport studies the ability of subjects to control blood flow and produce warmth or coldness in different parts of the body has been attributed to autogenic training.

Visualization can make it possible to picture physiological changes around the injured area. Physicians now use mental imagery not only to create a state of relaxation but also to facilitate and enhance the speed and effectiveness of the healing process. The basic technique of visualization is described in Chapter 5, but for enhancing performance rehabilitation the technique would correspond to the following:

- For visualization to be used effectively, a thorough diagnosis of the injury and an explanation of what has to happen internally to effect full recovery must be sought from both physicians and sport physiotherapists. Descriptions of the injury should be as vivid as possible (in colour).

- Next, an image is created that may be highly technical, or simple and imaginative. Perhaps the cricketer imagines a stress fracture site mending, the bone knitting together, the blood cells surrounding the site bringing new oxygen and replenishing and energizing the area, and all the 'bad blood' and waste products being removed. It is important that the healing image feels good and has personal significance to the individual.

Cricketers have nothing to lose and much to gain by experimenting in self-healing, self-talk and what has been called pain meditation by researchers in behavioural medicine.

(d) PSYCHOLOGICAL READINESS TO RETURN TO PLAY

An attitude that 'if the body is ready for a return to play, the mind is also' is both naïve and dangerous because it may lead to immediate re-injury or injury to another body part due to over-protection of the more seriously injured part. It could also cause lowered performance and possible loss of confidence. A concern therefore with only physical rehabilitation, such as absence of pain, full range of motion and full strength return, must be supplemented with psychological considerations.

Unfortunately, very little research has addressed the question of 'psychological readiness', to return to play. However, a holistic view of rehabilitation is advocated, involving:

- full discussions between the cricketer and the support staff (physicians, sport physiotherapist, sport psychology consultant, coach) about his or her readiness to return. In spite of physical evidence, diffident cricketers should never be returned to active duty, particularly if they believe they are not ready;

- a training programme of graduated levels of physical activity preceding the decision stage, comprising gentle-moderate-vigorous periods of cricket competition against different levels of opponents. This is essential for physical rehabilitation but it is also critical for restoring and sustaining the self-confidence (psychological rehabilitation) which is necessary for full readiness to return to active duty.

Summary

Currently the psychological care of injured cricketers has not been a priority amongst health care professionals and coaches, but it should be. Consequently the psychological impact of a serious injury is poorly understood. Cricketers themselves often appear to care less about such factors until they are personally affected (injured). With the advent of more sport psychology personnel available in the community this situation can change and should change.

An educational programme explaining the practical relevance of applied research which combines both physical and psychological intervention techniques is available from sport psychology consultants for both the medical and sport communities. Through this programme, the traumas associated with serious cricket injuries will be better understood from a holistic perspective and therefore better treated.

References

CHAPTER 1

Micheli, L. J. (1983) Overuse Injuries in Children's Sports: The Growth Factor, *Orthopedic Clinics of North America* 14(2), 337–57.

Stivens, D. (1955) The Batting Wizard from the City, *Australian Pageant*, P. R. Smith (ed.), Angus and Robertson, Melbourne.

CHAPTER 2

Davis, K., and B. Blanksby (1976) A Cinematographic Analysis of Fast Bowling in Cricket, *The Australian Journal for Health, Physical Education and Recreation*, March, 9–15.

Dawson, B., T. Ackland and C. Roberts (1984) A New Fitness Test for Team and Individual Sports, *Sports Coach* 8(2), 42–4.

Lillee, D. K. (1977) *The Art of Fast Bowling*, Collins, Sydney.

Willis, B. (1984) *Fast Bowling with Bob Willis*, Willow Books, London.

CHAPTER 3

Bradman, Sir D. (1958) *The Art of Cricket*, Hodder and Stoughton, London.

CHAPTER 6

Foster, D., B. Elliott, S. Gray and L. Herzberg (1984) Guidelines for the Fast Bowler, *Sports Coach* 7(4), 47–8.

Foster, D., D. John, B. Elliott, T. Ackland and K. Fitch (1989) Back Injuries to Fast Bowlers in Cricket: A Prospective Study, accepted *British Journal of Sports Medicine*.

Index

abdominal strength, 55, 73–5
aerobic, 4, 9–10, 54, 77–84, 86–7
air resistance, 38
anaerobic, 5, 10, 54, 84–5
asymmetry, 55–6

back strength, 55
batting, 36
bouncer, 42
bowling end, 34–5
bowling grip—*see* grip
bowling spell, 24–5, 53, 59

circuit training, 15–18
cognitive model, 90
co-ordination, 6, 32
corridor of uncertainty, 41
critical velocity, 38
cutting, 40–1

drag—air, 38
diet, 23

equipment, 23

Fartlek training, 9
fat, 5
fielding, 35–6
fitness:
  assessment, 6, 7
  components, 4
  development, 6–12
flexibility, 6, 10–15, 70–2
fluid intake, 22–3
follow-through, 33
food, 23
foot arches, 55
force, 58

grip:
  basic, 28

inswing, 39–40
leg-cutter, 41
off-cutter, 41
one-finger, 42
outswing, 39
palm ball, 42
slow ball, 42

hamstrings, 54
health, 23

injuries:
  back, 61–5
  disc, 63–4
  lower-limb, 65–6
  physical demand, 57–9
  prevention, 54–9
  psychological response, 88–9
  psycho-social causes, 87
  spondylolisthesis, 63
  stress fracture, 61–3
  techniques, 56–7
  trunk, 66–7
  upper-limb, 67

landing force, 58
late swing, 38
lordosis, 56

muscular endurance, 5

neuromuscular co-ordination, 85

outswinger, 39
overuse, 57–9

physical attributes, 54
plyometrics, 84
pool running, 80–4
posture, 6, 55–6

psychology:
   anxiety, 47–9
   arousal, 47, 52–3
   closed skill, 43
   communication, 53
   concentration, 46–7
   confidence, 50–2
   game strategies, 52–3
   goal setting, 44–6
   mental consistency, 50–2
   mental skills, 44–6
   mental training—*see* visualization
   mental toughness, 47
   pressure, 47–9
   psycho-social causes of injury, 87
   rehabilitation, 89–93
   response to injury, 88–9
   stress management, 47–9
   visualization skills, 49–50

quadriceps, 54

rehabilitation:
   abdominal strength, 73–5
   aerobic fitness, 77–84
   aerobics, 86–7
   anaerobic, 84–5
   back strength, 76
   diagnosis, 68–9
   flexibility, 70–2
   initial treatment, 68–9
   neuromuscular co-ordination, 85
   plyometrics, 84
   pool running, 80–4
   psychological aspects, 89–92

   return to play, 92–3
   swimming, 77–80

scoliosis, 56
seaming, 40–1
skinfold, 5
slow ball, 41–2
specificity, 7
spell of bowling, 24–5
sprint training, 9–10
spondylolisthesis, 63
strength, 5, 15–22, 54, 55, 73–6
stride length, 31
stress fracture, 61–3
swimming, 77–80
swing bowling, 37–42

training, 7–22
technique:
   back-foot landing, 27
   delivery body position, 33
   delivery stride, 28–32
   follow-through, 33
   front-elbow movement, 31
   run up, 27
   shoulder alignment, 28–31
   stride length, 31

visualization, 49–50

warm-up, 22
weight lifting, 19–22
wobble board, 85

yorker, 42